CRC Handbook of Oligosaccharides

Volume III
Higher Oligosaccharides

Authors

András Lipták, D.Sc.
Professor
Institute of Biochemistry
Lajos Kossuth University
Debrecen, Hungary

Zoltán Szurmai, Ph.D.
Scientific Fellow
Institute of Biochemistry
Lajos Kossuth University
Debrecen, Hungary

Péter Fügedi, Ph.D.
Scientific Fellow
Institute of Biochemistry
Lajos Kossuth University
Debrecen, Hungary

János Harangi, Ph.D.
Scientific Fellow
Institute of Biochemistry
Lajos Kossuth University
Debrecen, Hungary

CRC Press
Boca Raton Ann Arbor Boston London

Library of Congress Cataloging-in-Publication Data
(Revised for vol. 3)

Lipták, András, D.Sc.
CRC handbook of oligosaccharides.

Cover title: Handbook of oligosaccharides.
Vol. 2.: authors, András Lipták . . . [et al.].
Includes bibliographical references and indexes.
Contents: v. 1. Disaccharides — v. 2. Trisaccharides — v. 3. Higher oligosaccharides.
1. Oligosaccharides—Synthesis—Handbooks, manuals, etc. I. Fügedi, Péter, Ph.D. II. Szurmai, Zoltán. III. Handbook of oligosaccharides.
[DNLM: 1. Oligosaccharides. QU 83 H236]
QP702.044H36 1990 547.7′815 90-1755
ISBN 0-8493-2903-5

Direct all inquiries to CRC Press, Inc., 2000 Corporate Blvd., N.W., Boca Raton, Florida 33431.

International Standard Book Number 0-8493-2903-5

Library of Congress Card Number 90-1755
Printed in the United States

INTRODUCTION

The number of free oligosaccharides occurring in nature is relatively low, but their derivatives in the form of, for example, plant glycosides, antibiotics, glycolipids, and mainly glycoproteins are indeed widespread. Especially in the case of the last two groups, to clear up their biological roles, intense research has begun not only in the circles of biochemists and immunologists but also among the synthetic carbohydrate chemists.

The improvement in the efficiency of isolation techniques and high performance spectroscopic methods has enabled unprecedented development in the area of oligosaccharides. The extreme variety of structures and in some cases the miniscule amounts of isolated materials involved have presented new challenges for synthetic chemists; new synthetic methods had to be developed to ensure sufficient quantities for biological investigations and to enable the production of varied structures. Extraordinary progress has been achieved in two areas of synthesis. First, with the aid of new blocking strategies a wide range of partially protected mono- and oligosaccharides have become available. Secondly, with the exploitation of the more detailed mechanism of glycosylation reactions, new methods appeared which were suitable for the production of 1,2-cis and 1,2-trans glycosides. As a result, in the past 10 years an extremely wide range of oligosaccharide syntheses could be achieved.

Although during the past decade a number of excellent reviews have been published to assist in the survey of the newest developments in the field, there has been no book that would provide a survey of the structures of the oligosaccharides synthesized to date and the details of the applied methods. We felt, on the basis of our own experience, that such a book would represent a great help to the carbohydrate chemist and at the same time to all scientists who deal in some ways with oligosaccharides.

Our collection is meant to meet the needs, first of all, of those dealing in synthetic carbohydrate chemistry and beyond merely listing the syntheses, also showing the route of the synthesis. Thus, not only are those oligosaccharides presented which were prepared in their free form, but also those produced in a protected form. In the planned series of three volumes can be found all of the oligosaccharides synthesized between 1960 and 1986. We present in tabular form the aglycon, the glycosyl donor, the reaction conditions applied (solvent, promoter, temperature), and the structure of the isolated product. In the case of disaccharides the names of the reactants and the products are given, while in the case of the trisaccharides and higher oligomers schematic figures provide quick and easy information concerning the entire synthesis.

THE AUTHORS

András Lipták, Ph.D., D.Sci., is Professor of Biochemistry in the Department of Biology at Kossuth Lajos University, Debrecen, Hungary. He received his M.Sc. degree from the same university in 1961, graduating with highest honors. In 1968 he received his Ph.D. in organic chemistry and in 1983 he was honored by the Hungarian Academy of Sciences as D.Sc.

Professor Lipták held an Alexander von Humboldt Fellowship in Munich in 1971-1972 and he spent nearly 2 years at the National Institutes of Health (Bethesda, Maryland). He is an author of over 110 articles and has presented his scientific results at numerous international meetings. He is the member of the Editorial Board of the *Journal of Carbohydrate Chemistry*. His main research interest covers the selective protecting of carbohydrates and synthesis of complex oligosaccharides. In 1989 the Hungarian Academy of Sciences awarded him with the Zemplén Géza Prize. In 1990 Professor Lipták was elected to be the Rector of Kossuth Lajos University and also was selected a member of the Hungarian Academy of Sciences.

Zoltán Szurmai, Ph.D., is a scientific fellow at the Institute of Biochemistry, Kossuth Lajos University in Debrecen, Hungary. He graduated from Kossuth Lajos University in 1977 as a chemist. Dr. Szurmai received his Ph.D. degree from the same university in 1982. He then completed a 1-year fellowship at the Gerontology Research Center, National Institutes of Health, National Institute on Aging, in Baltimore, Maryland. He worked for Ruhr University in 1988 and 1989 for a short time in West Germany. His major research interest is the chemical syntheses of oligosaccharides. He has 17 articles published in international journals. He has 18 articles in international journals. He is presently working in Orsay, France at C.N.R.S. Endotoxines, Institut de Biochimie, Université de Paris Sud.

Péter Fügedi, Ph.D., is currently a visiting scientist at Glycomed Inc. in Alameda, California. He received his M.Sc. degree in chemistry at Kossuth Lajos University, Debrecen, Hungary in 1975, and his Ph.D. degree in 1978 from the same university. He got his habilitation from the Hungarian Academy of Sciences in 1989. Being affiliated at the Institute of Biochemistry of Lajos Kossuth University, he had postdoctoral experience at the Laboratory of Structural Biochemistry in Orleans, France in 1978-1979, and was a visiting scientist at the Department of Organic Chemistry, Arrhenius Laboratory, University of Stockholm, Sweden in 1984-1985. He is an author of over 30 articles and presented his results at numerous international meetings. His major research interest is synthetic carbohydrate chemistry, with emphasis on glycosylation methods and the synthesis of oligosaccharides.

János Harangi, Ph.D., is a scientific fellow at the Institute of Biochemistry, Kossuth Lajos University, Debrecen, Hungary. He graduated from Kossuth Lajos University in 1974 as a chemist. In 1980 he received his Ph.D. in biochemistry from the same university. He worked at Munich University and Ruhr University in West Germany several times. His main research interest is the structure investigation of sugar derivatives by nuclear magnetic resonance spectroscopy and separation techniques in carbohydrate chemistry. He has 30 articles published in international journals. He is now working for Hewlett-Packard Vienna as a sales representative and customer trainer.

ADVISORY BOARD

HOW TO USE THE BOOK

Handbook of Oligosaccharides, Volume III contains all the chemical syntheses of higher oligosaccharides published between 1960—1986 in the following sequence:

Tetrasaccharides
Pentasaccharides
Hexasaccharides
Heptasaccharides
Octasaccharides
Nonasaccharides
Undecasaccharides

In the hierarchy of the monosaccharide building blocks the following order is used: Xylose, Altrose, Glucose, Mannose, Idose, Galactose, Talose, KDO, Neuraminic Acid, Deoxy Sugars, Fructose.

The D-sugars precede the L-sugars. The uronic acids follow immediately after the parent sugars. Some deoxy- and aminodeoxy-sugars like 2-amino-2-deoxy-D-glucose, -D-mannose, and -D-galactose can be found immediately after uronic acids. Among the 6-deoxy-sugars, the rhamnose (6-deoxy-mannose) and fucose (6-deoxy-galactose) have the same privilege. All other amino-sugars, deoxy-sugars, and deoxy-halogenosugars are itemized after the neuraminic acid.

Concerning the ring size, the furanosides precede the pyranosides. The α-anomers precede the β-anomers. The sequence of the bond-types is the following: (1→2); (1→3); (1→4); i.e., instead of the full name of oligosaccharides the abbreviated name is given, e.g., α-Tyvp-(1→3)-α-D-Manp-(1→4)-α-L-Rhap-(1→3)-D-Gal instead of *O*-(α-Tyvelopyranosyl)-*O*-(α-D-mannopyranosyl)-(1→4)-*O*-(α-L-rhamnopyranosyl)-(1→3)-D-galactose. Linear oligosaccharides precede branched ones when they have the same sugar at the reducing end.

If an oligosaccharide is not listed it was most probably not prepared synthetically in that period. Under an individual oligosaccharide entry you can find the following information: the abbreviated name of the oligosaccharide; physical data of the free oligosaccharide (m.p. $/\alpha/_D$ value) if it is prepared.

In some cases if the most important glycosides of the oligosaccharide are known, one can find symbols of these; e.g., β-p-(1→OMe) means that the β-methyl glycoside of the oligosaccharide was prepared and the reducing end was in pyranoside form; physical data of the important glycosides. One can also get information about the glycosylation reactions: there are schematic figures of the aglycons on the far left and the glycosyl donors left of center; in the middle one can see the reaction conditions (catalyst, solvent). Different reaction conditions are numbered with roman numerals.

We used abbreviations for catalysts and solvents (see list of abbreviations). Schematic figures of the products are located on the far right or at the bottom of the page. Underneath the figures one can find data of the products (e.g. yield, m.p.,$/\alpha/_D$ value). C-13 means that ^{13}C-NMR data are available in the original paper. We used abbreviations for the protecting groups in schematic figures (see list of abbreviations).

ABBREVIATIONS

Sugars

Abep	=	3,6-Dideoxy-D-*xylo*-hexopyranosyl
Alt	=	Altrose
Ascp	=	3,6-Dideoxy-L-*arabino*-hexopyranosyl
Fuc	=	Fucose
Fru	=	Fructose
Gal	=	Galactose
GalN	=	2-Amino-2-deoxy-galactose
GalNAc	=	2-Acetamido-2-deoxy-galactose
Glc	=	Glucose
GlcA	=	Glucuronic acid
GlcN	=	2-Amino-2-deoxy-glucose
GlcNAc	=	2-Acetamido-2-deoxy-glucose
Hex	=	Hexose
IdopA	=	Idopyranosyluronic acid
KDO	=	3-Deoxy-D-*manno*-2-octulopyranosic acid
Man	=	Mannose
ManN	=	2-Amino-2-deoxy-mannose
ManNAc	=	2-Acetamido-2-deoxy-mannose
Neu5Ac	=	*N*-Acetyl-neuraminic acid
Parp	=	3,6-Dideoxy-D-*ribo*-hexopyranosyl
Rha	=	Rhamnose
Tal	=	Talose
Tyvp	=	3,6-Dideoxy-D-*arabino*-hexopyranosyl
Xyl	=	Xylose

Protecting groups

Ac	=	Acetyl
All	=	Allyl
Bz	=	Benzoyl
Bzl	=	Benzyl
Bu	=	Butyl
tBu	=	*tert*-Butyl
$tBuMe_2Si$	=	*tert*-Butyldimethylsilyl
$tBuPh_2Si$	=	*tert*-Butyldiphenylsilyl
Cbz	=	Carbobenzyloxy
$C(Me)_2$	=	Isopropylidene
Dcp	=	2,3-Diphenyl-2-cyclopropenyl
ECO	=	8-Ethoxycarbonyloctyl
Et	=	Ethyl
Lev	=	Levulinoyl
MBA	=	Monobromoacetyl
MCA	=	Monochloroacetyl
MCO	=	8-Methoxycarbonyloctyl
Me	=	Methyl
NBz	=	*p*-Nitrobenzoyl
Ph	=	Phenyl
Phth	=	Phthalimido
PNP	=	*p*-Nitrophenyl

Pr = Propyl
TCA = Trichloroacetyl
Ts = *p*-Toluenesulfonyl
Tr = Triphenylmethyl
Tres = 2,2,2-Trifluoroethanesulfonyl

Reagents and solvents

Ac_2O = Acetic anhydride
AgOTf = Silver trifluoromethanesulfonate
AgSi = Silver silicate
Bu_4NBr = Tetrabutylammonium bromide
$2,6(tBu)_2Py$ = 2,6-Di-*tert*-butylpyridine
s-Coll = Collidine (2,4,6-Trimethylpyridine)
DMF = *N,N*-Dimethylformamide
$DpcpClO_4$ = 2,3-Diphenyl-2-cyclopropen-1-ylium perchlorate
Et_3N = Triethylamine
Et_4NBr = Tetraethylammonium bromide
Et_2O = Diethyl ether
Lut = Lutidine (2,6-Dimethylpyridine)
MeCN = Acetonitrile
$MeNO_2$ = Nitromethane
MeOH = Methyl alcohol
MeOTf = Methyl trifluoromethanesulfonate
NaOMe = Sodium methoxide
pNBCl = *p*-Nitrobenzenesulfonyl chloride
NIS = *N*-Iodosuccinimide
PdC = Palladium on carbon
$(iPr)_2EtN$ = *N,N*-Diisopropylethylamine
Py = Pyridine
$PyH^+ClO_4^-$ = Pyridinium perchlorate
TMSOTf = Trimethlylsilyl trifluoromethanesulfonate
TMU = 1,1,3,3-Tetramethylurea
$TrClO_4$ = Triphenylmethylium perchlorate
pTSA = *p*-Toluenesulfonic acid

TETRASACCHARIDES

Having monosaccharide units at the reducing end in the following sequence:

D-Xyl
D-Glc
D-GlcN (D-GlcNAc)
D-Man
D-ManN (D-ManNAc)
L-Rha
D-Gal
D-GalN (D-GalNAc)
3-Deoxy-3-fluoro-D-Gal
6-Deoxy-L-Tal
2,6-Dideoxy-D-*ribo*-Hexp
D-Fru

β-D-Xylp-(1→3)-β-D-Xylp-(1→3)-β-D-Xylp-(1→3)-D-Xyl

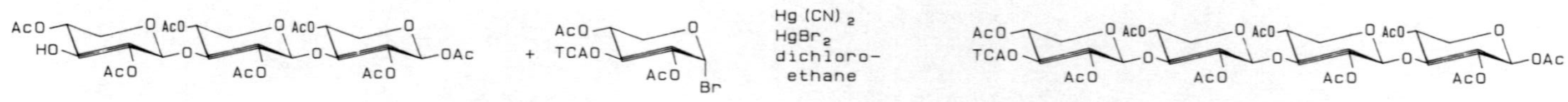

m.p. 160 °C; $[\alpha]_D$ −76.8 (chloroform); C-13; Ref.: 1

α-D-Xylp-(1→4)-β-D-Xylp-(1→4)-β-D-Xylp-(1→4)-D-Xyl

β-p-(1→OMe) acetate; amorphous solid; $[\alpha]_D$ −38 (chloroform); C-13; Ref.: 2

β-acetate; m.p. 137–138 °C; $[\alpha]_D$ −29.3 (chloroform); C-13; Ref.: 3

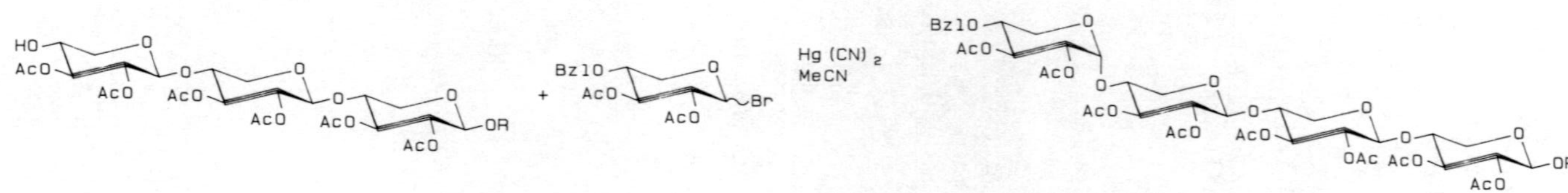

R=Me; 30.2%; amorphous solid
$[\alpha]_D$ −37 (chloroform); C-13; Ref.: 2

R=Ac; 32.5%; colourless foam
$[\alpha]_D$ −31.1 (chloroform); C-13; Ref.: 3

β-D-Xylp-(1→4)-β-D-Xylp-(1→4)-β-D-Xylp-(1→4)-D-Xyl

m.p. 223.5–225.5 °C; $[\alpha]_D$ −58 (water); C-13; Ref.: 3

β-acetate; m.p. 200–201 °C; 210–211 °C (dimorphous); $[\alpha]_D$ −95.9 (chloroform); Ref.: 3

β-p-(1-OMe); m.p. 238–239 °C; $[\alpha]_D$ −88 (water); C-13; Ref.: 2, 4

β-p-(1-OMe) acetate; m.p. 245–246 °C; $[\alpha]_D$ −111 (chloroform); C-13; Ref.: 2

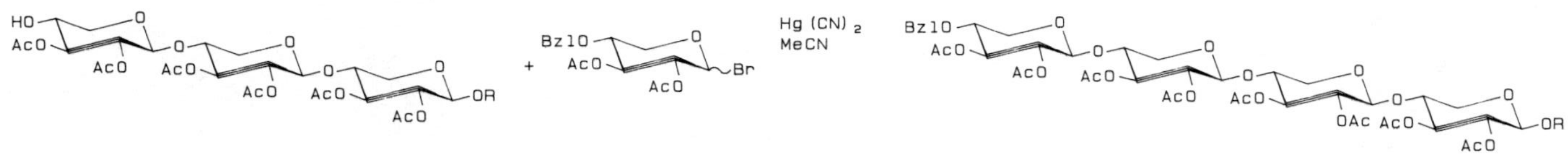

R=Me; 47.8%; m.p. 198–199 °C;
$[\alpha]_D$ −106 (chloroform); C-13; Ref.: 2, 4

R=Ac; 51%; m.p. 188.5–191 °C;
$[\alpha]_D$ −91.5 (chloroform); C-13; Ref.: 3, 4

β-D-Xylp-(1→3)
β-D-Xylp-(1→4) 〉 β-D-Xylp-(1→4)-D-Xyl

β-p-(1→OMe); colourless solid foam; $[\alpha]_D$ -87.7 (water); Ref.: 5

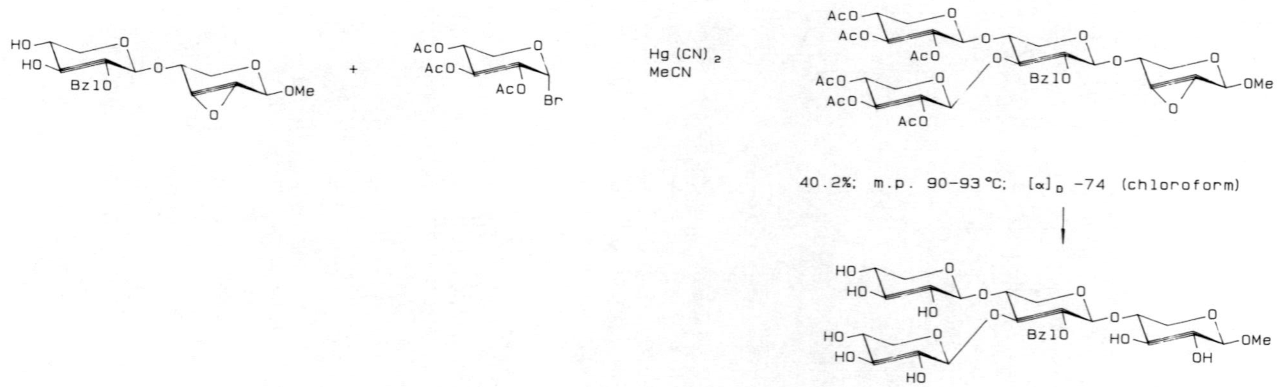

40.2%; m.p. 90-93 °C; $[\alpha]_D$ -74 (chloroform)

m.p. 260-261 °C; $[\alpha]_D$ -87.4 (water); Ref.: 5

α-D-Glcp-(1→4)-α-D-Glcp-(1→4)-α-D-Glcp-(1→4)-D-Glc

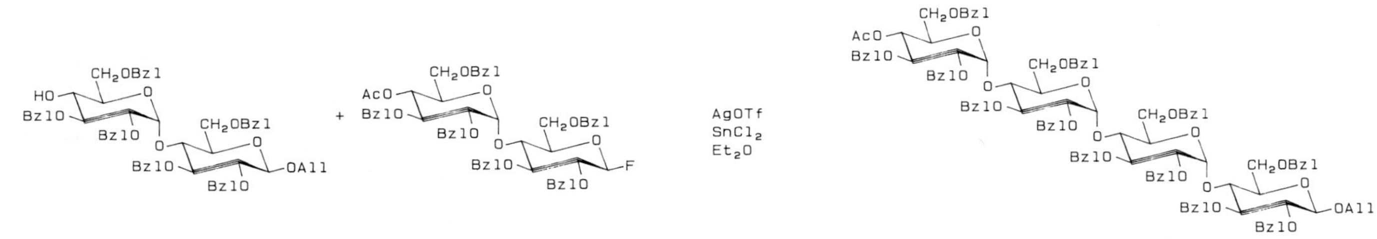

$[\alpha]_D$ +58.2 (chloroform); C-13; Ref.: 6

α-D-Glcp-(1→4)-β-D-Glcp-(1→4)-α-D-Glcp-(1→4)-D-Glc

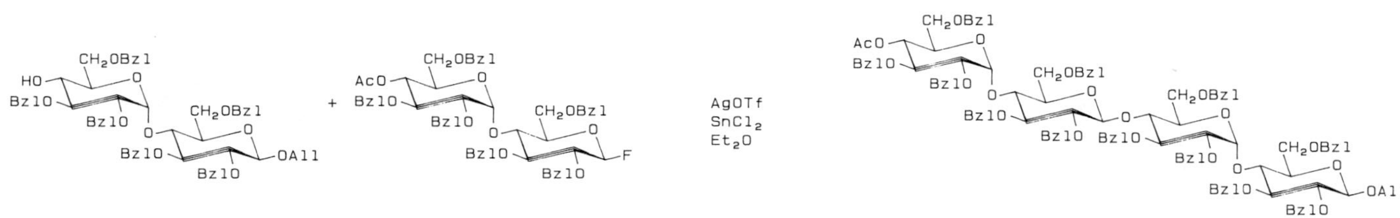

$[\alpha]_D$ +41.4 (chloroform); C-13; Ref.: 6

α-D-Glcp-(1→4)-β-D-Glcp-(1→6)-α-D-Glcp-(1→4)-D-Glc

$[\alpha]_D$ +97.5 (water); Ref.: 7

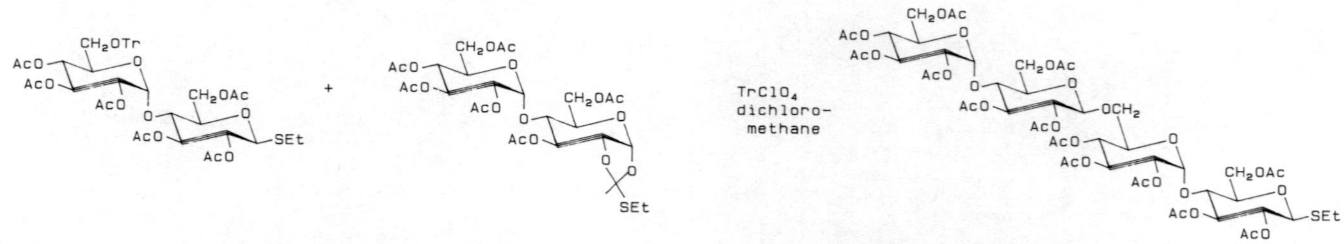

60%; m.p. 192–194 °C; $[\alpha]_D$ +53 (chloroform); Ref.: 7

α-D-Glcp-(1→6)-α-D-Glcp-(1→6)-α-D-Glcp-(1→6)-D-Glc

hygroscopic glass; $[\alpha]_D$ +148 (water); Ref.: 8, 9

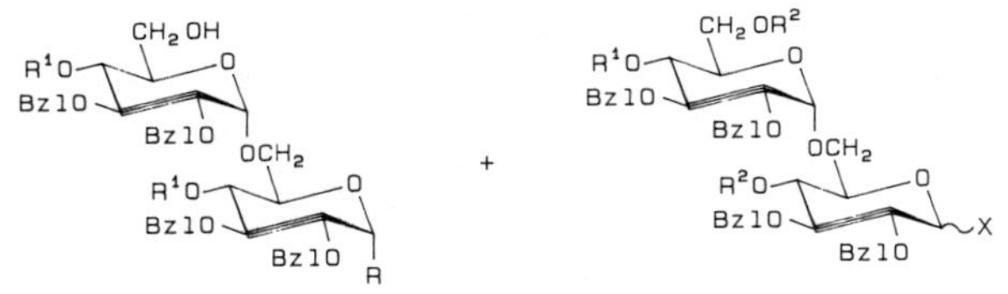

I Lut, $MeNO_2$

II $AgClO_4$, $SnCl_2$, Et_2O

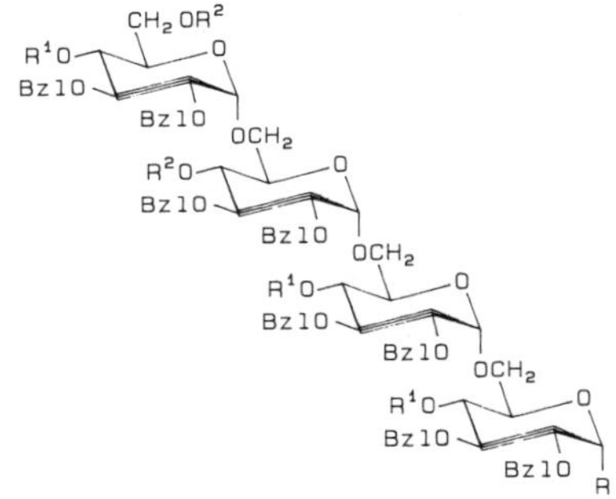

I R=SEt; R^1=Bzl; R^2=NBz; X=Br
42–59%; $[\alpha]_D$ +108 (chloroform); Ref.: 8, 9

II R=SPh; R^1=Ac; R^2=Ph_2tBuSi; X=F; 70%; Ref.: 10

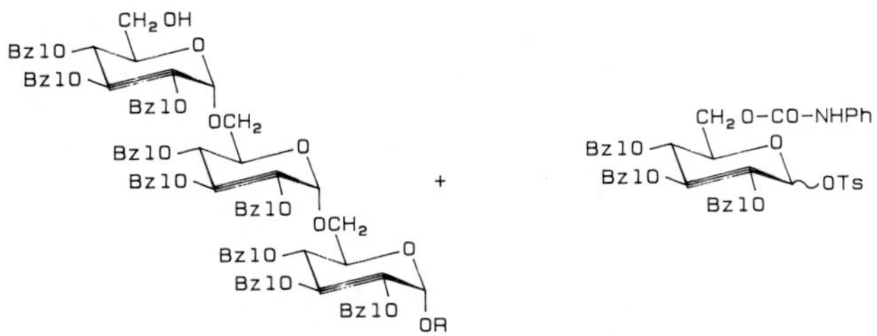

III Et_2O

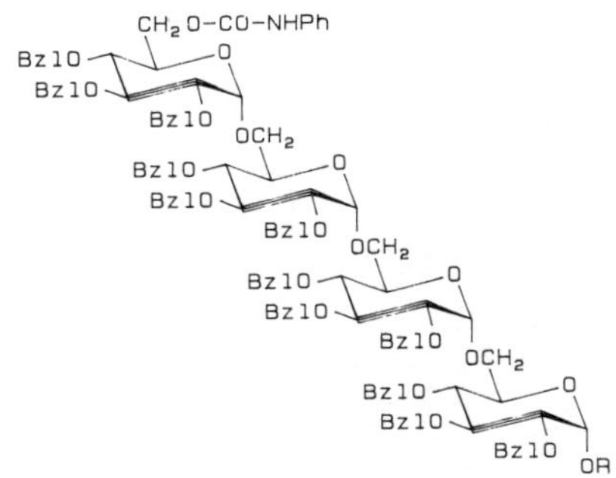

III R=Me; 85%; $[\alpha]_D$ +77.1 (chloroform); Ref.: 11, 12

R=$(CH)_2$-Ph-NHTs; 83%; $[\alpha]_D$ +74.9 (chloroform); Ref.: 13

β-D-Glcp-(1→2)-β-D-Glcp-(1→2)-β-D-Glcp-(1→2)-D-Glc

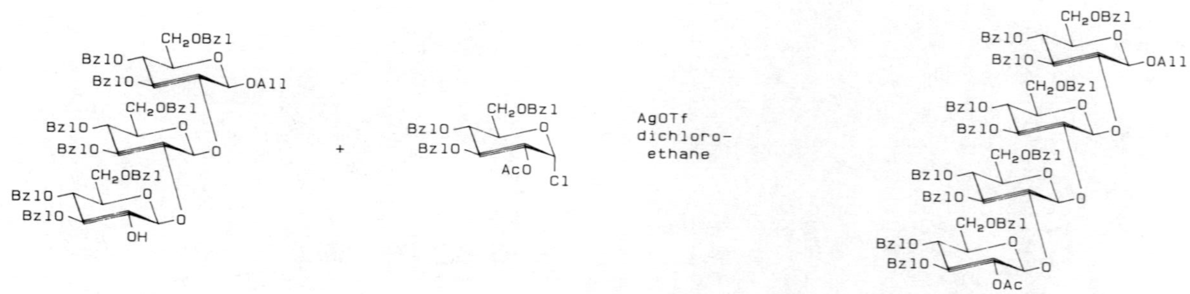

75%; $[\alpha]_D$ -0.63 (chloroform); C-13; Ref.: 14

β-D-Glcp-(1→6)-β-D-Glcp-(1→3)-β-D-Glcp-(1→3)-D-Glc

amorphous; $[\alpha]_D$ -17 (water); Ref.: 15

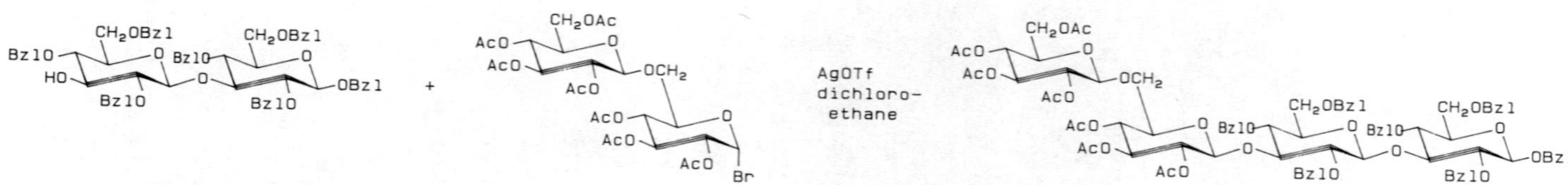

80%; amorphous; $[\alpha]_D$ -2 (chloroform); C-13; Ref.: 15

β-D-Glcp-(1→4)-β-D-Glcp-(1→4)-β-D-Glcp-(1→3)-D-Glc

m.p. 257-260 °C (dec.); $[\alpha]_D$ +12 → +9 (water); C-13; Ref.: 16

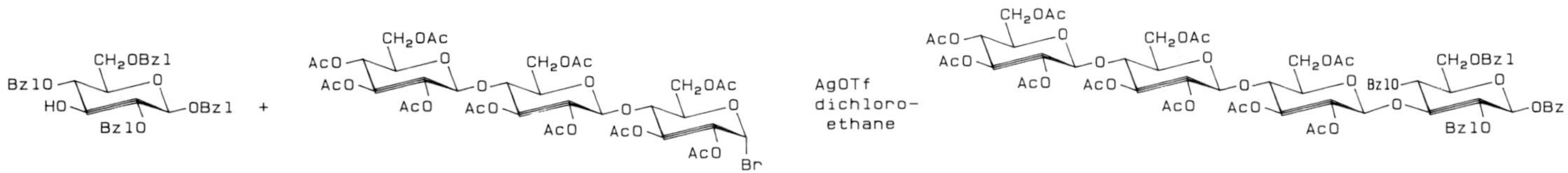

75%; $[\alpha]_D$ -14.5 (chloroform); C-13; Ref.: 16

β-D-Glcp-(1→3)
β-D-Glcp-(1→6) 〉β-D-Glcp-(1→3)-D-Glc

m.p. 199-200 °C; $[\alpha]_D$ -2 → -6 (water); Ref.: 15

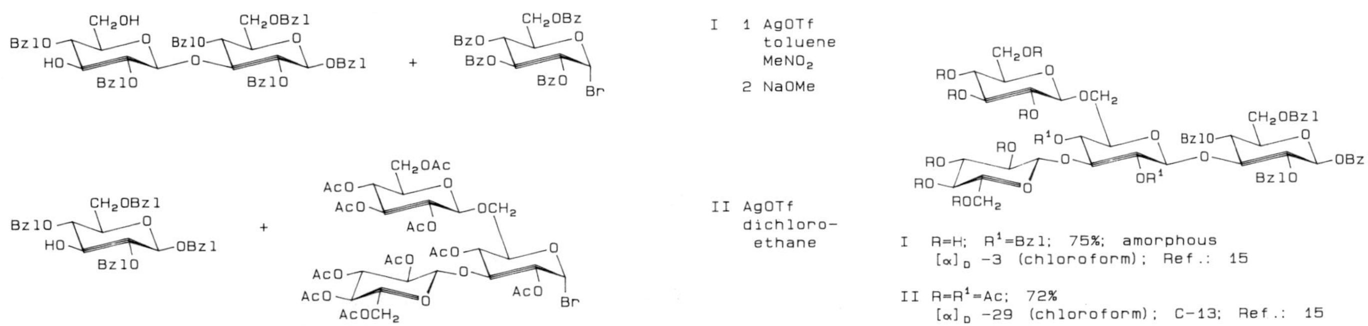

I R=H; R^1=Bzl; 75%; amorphous
$[\alpha]_D$ -3 (chloroform); Ref.: 15

II R=R^1=Ac; 72%
$[\alpha]_D$ -29 (chloroform); C-13; Ref.: 15

β-D-Glcp-(1→4)-β-D-Glcp-(1→3)-β-D-Glcp-(1→4)-D-Glc

m.p. 224–227 °C; $[\alpha]_D$ +21 (DMSO); C-13; Ref.: 16

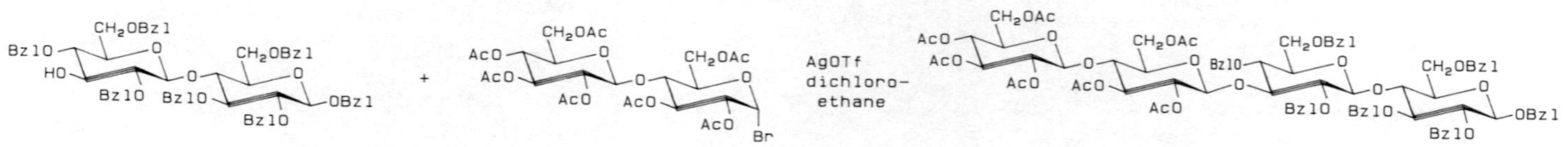

84%; amorphous; $[\alpha]_D$ −10 (chloroform); C-13; Ref.: 16

β-D-Glcp-(1→3)-β-D-Glcp-(1→4)-β-D-Glcp-(1→4)-D-Glc

m.p. 220–223 °C; (dec.) $[\alpha]_D$ +13 → +10 (water); Ref.: 16

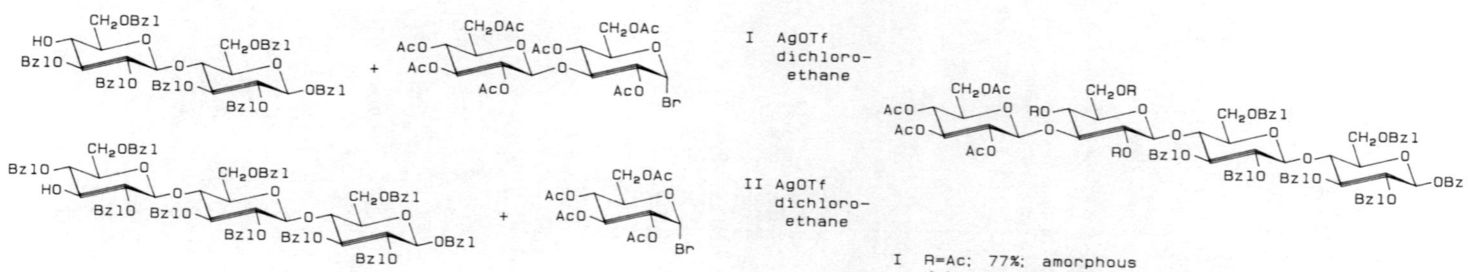

I R=Ac; 77%; amorphous
$[\alpha]_D$ −21 (chloroform); C-13; Ref.: 16

II R=Bzl; 83%; amorphous
$[\alpha]_D$ −1 (chloroform); C-13; Ref.: 16

β-D-Glcp-(1→4)-β-D-Glcp-(1→4)-β-D-Glcp-(1→4)-D-Glc

m.p. 251-253 °C; (dec.) $[\alpha]_D$ +7.1 → +17.1 (water); Ref.: 17

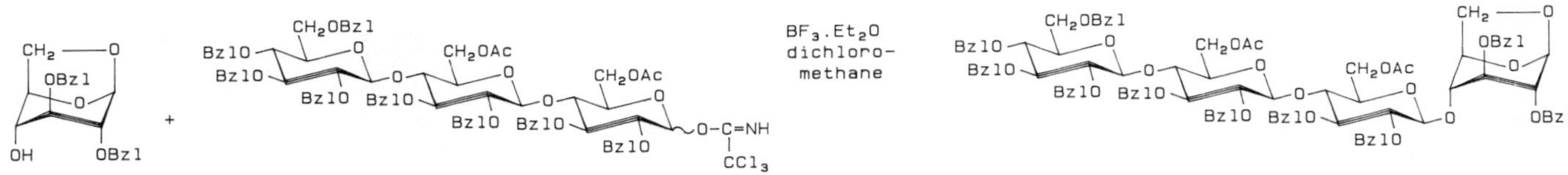

40%; Ref.: 18

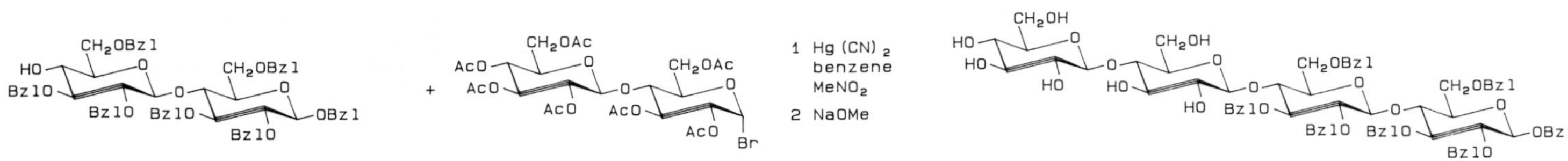

69%; amorphous powder;
$[\alpha]_D$ +5.2 (chloroform); Ref.: 17

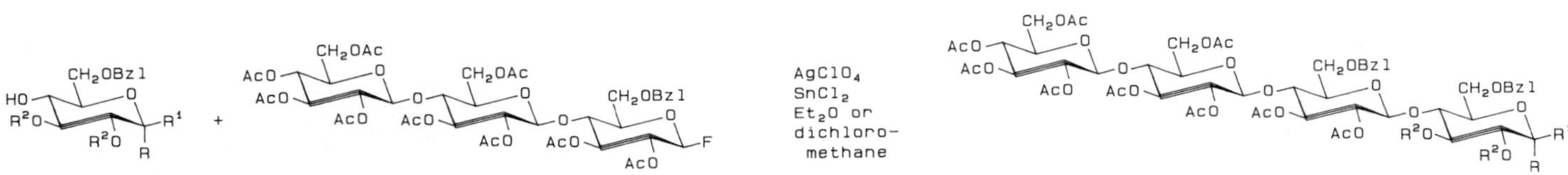

R^1=SPh; R=H; R^2=Ac; No data; Ref.: 19

R=O (H, OSi-tBuPh$_2$); R^1=H; R^2=Bzl; 74%; Ref.: 19

R=O (Ph$_2$tBu-SiO, H); R^1=H; R^2=Bzl; 74%; Ref.: 19

β-D-Glcp-(1→4) / β-D-Glcp-(1→6) 〉 β-D-Glcp-(1→4)-D-Glc

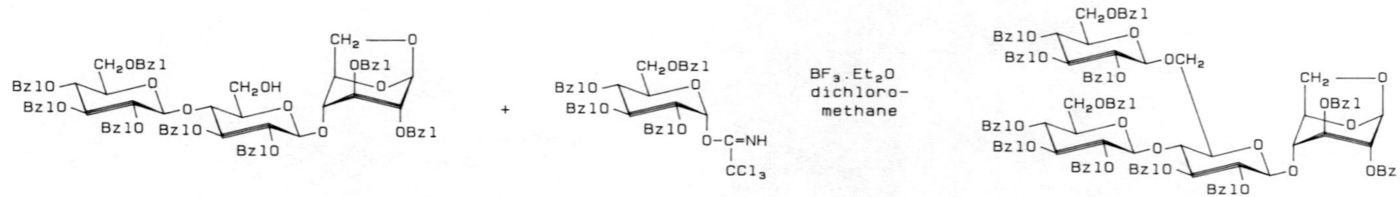

70%; Ref.: 18

β-D-Glcp-(1→3)-β-D-Glcp-(1→3)-β-D-Glcp-(1→6)-D-Glc

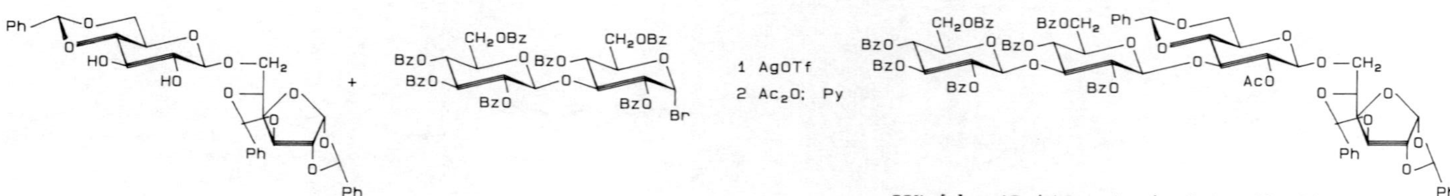

69%; $[\alpha]_D$ −16 (chloroform); Ref.: 20, 21

β-D-Glcp-(1→6)-β-D-Glcp-(1→6)-β-D-Glcp-(1→6)-D-Glc

amorphous; m.p. 164–171 °C; $[\alpha]_D$ -6.5 (water); Ref.: 22

amorphous powder; $[\alpha]_D$ -15.5 (water); Ref.: 23

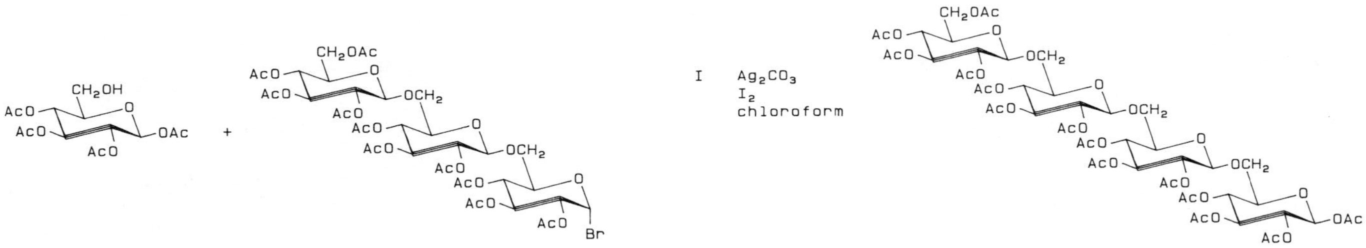

17%; m.p. 212–213 °C; $[\alpha]_D$ -10.7 (chloroform); Ref.: 22

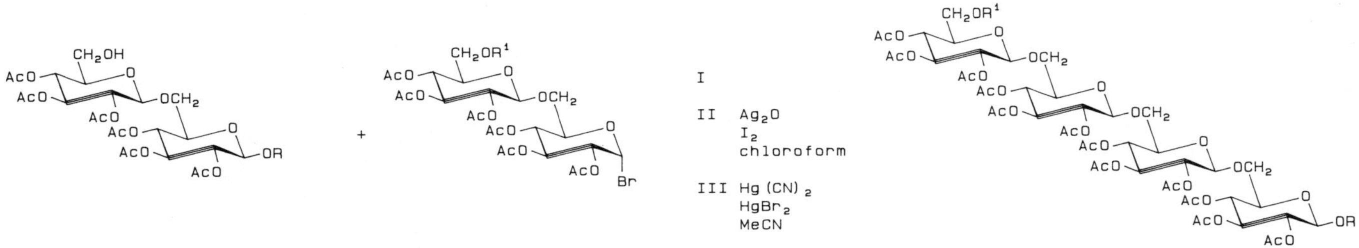

I R=R[1]=Ac; 31%; 212–213 °C;
$[\alpha]_D$ -10.7 (chloroform); Ref.: 22

III R=R[1]=Ac; 60%; 211–213 °C;
$[\alpha]_D$ -10 (chloroform); Ref.: 23

II R=-Ph-OAc; R[1]=Ac; m.p. 147–152 °C; $[\alpha]_D$ -49; Ref.: 24

III R=Ac; R[1]=TCA; 75%; m.p. 176–177 °C;
$[\alpha]_D$ -6 (chloroform); Ref.: 23

β-D-Glcp-(1→3)
β-D-Glcp-(1→6) 〉β-D-Glcp-(1→6)-D-Glc

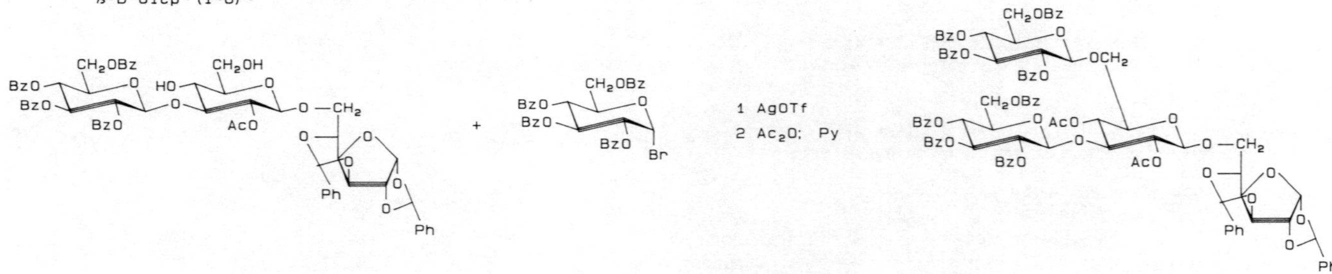

87%; Ref.: 20

endo $[\alpha]_D$ −8 (chloroform); Ref.: 21
exo $[\alpha]_D$ −4 (chloroform); Ref.: 21

→6)-[β-D-Glcp]$_4$-(1→ (cyclic)

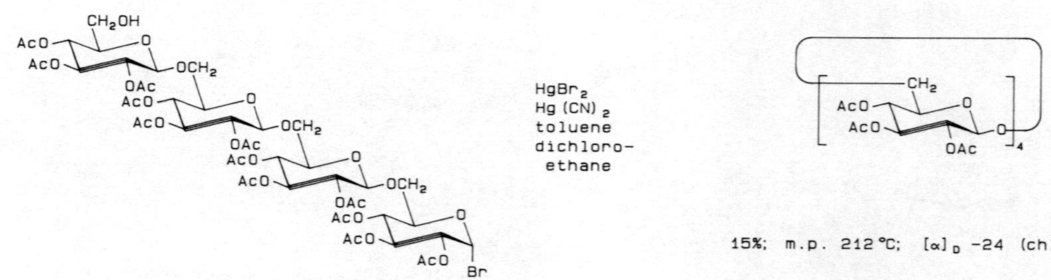

15%; m.p. 212 °C; $[\alpha]_D$ −24 (chloroform); C-13; Ref.: 25

α-D-Glcp-(1→6)-α-D-Glcp-(1→4) \
α-D-Glcp-(1→6) / D-Glc

$[\alpha]_D$ +152.4 (water); Ref.: 26

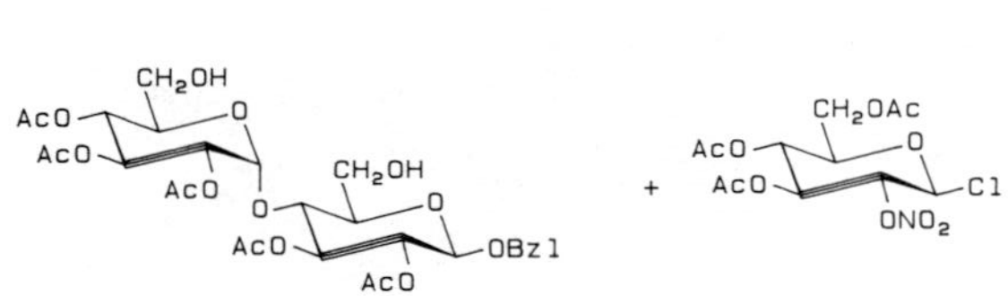

1 $AgClO_4$
Ag_2CO_3
benzene
2 H_2; PdC
3 NaOMe

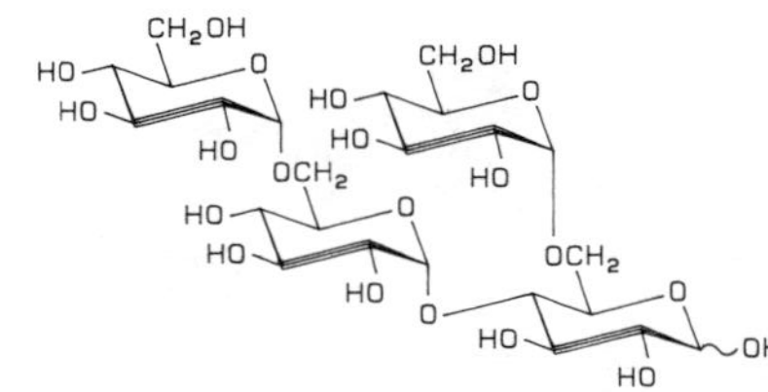

13%; Ref.: 26

β-D-Glcp-(1→6)-α-D-Glcp-(1→4) \
β-D-Glcp-(1→6) / D-Glc

amorphous; $[\alpha]_D$ +40 (water); Ref.: 26, 27

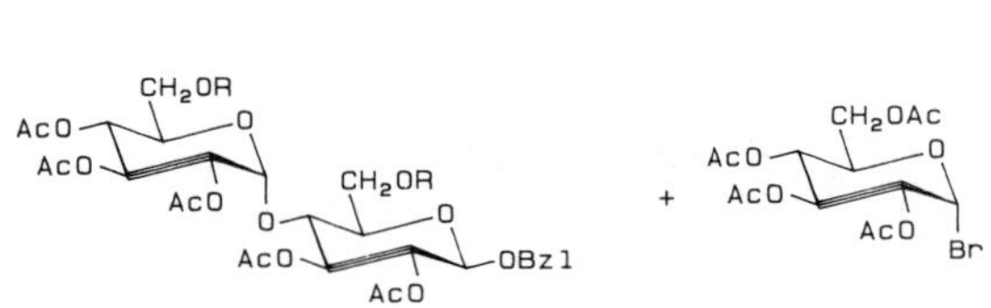

I R=H
1 Ag_2O; I_2
chloroform
2 H_2; PdC
3 NaOMe

II R=Tr
1 $AgClO_4$
$MeNO_2$
2 H_2; PdC
3 NaOMe

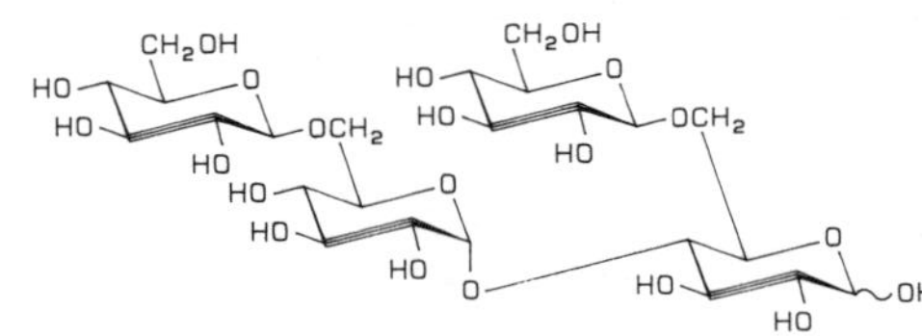

I 6%; Ref.: 26, 27

II 5%; Ref.: 26, 27

β-D-Glcp-(1→3)-β-D-Glcp-(1→3)
β-D-Glcp-(1→6)
⟩D-Glc

[α]$_D$ −0.3 (water); C-13; Ref.: 28

[α]$_D$ −3 (water); C-13; Ref.: 15

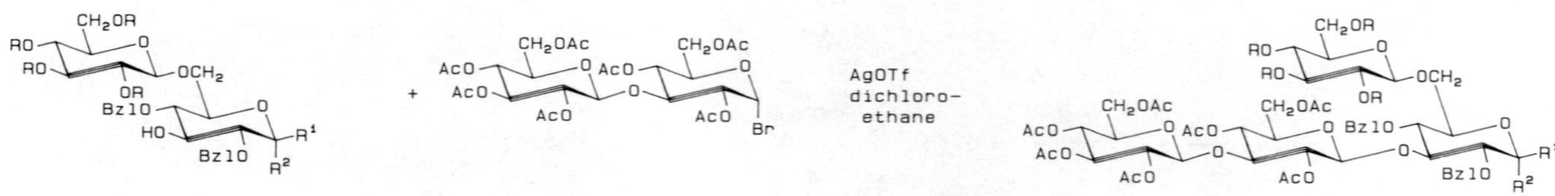

R=Ac; R^1=H; R^2=OBzl; 81%
[α]$_D$ +1.8 (chloroform); Ref.: 28

R=Bz; R^1=OBzl; R^2=H; 78%
[α]$_D$ −4 (chloroform); C-13; Ref.: 15

α-D-Galp-(1→6)-β-D-Glcp-(1→6)-α-D-Galp-(1→6)-D-Glc

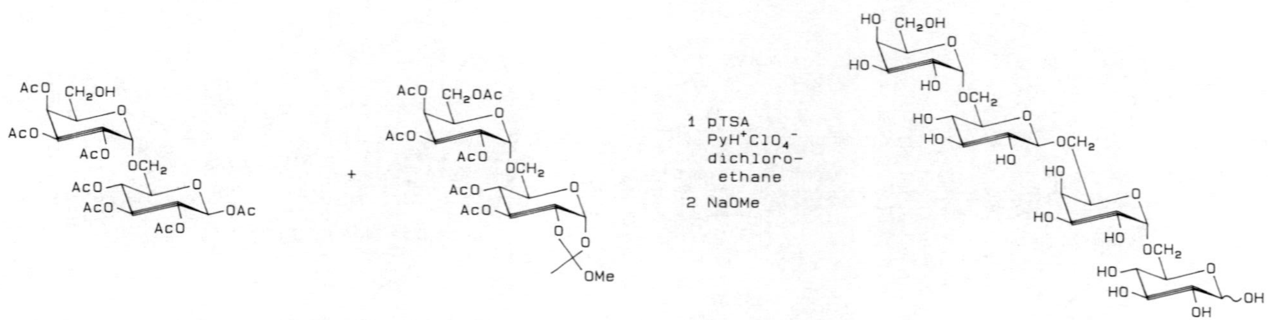

8%; Ref.: 29

β-D-Galp-(1→6)-α-D-Galp-(1→6)-α-D-Galp-(1→6)-D-Glc

β-p-(1→OMe); white powder; $[\alpha]_D$ +89.5 (water); Ref.: 30

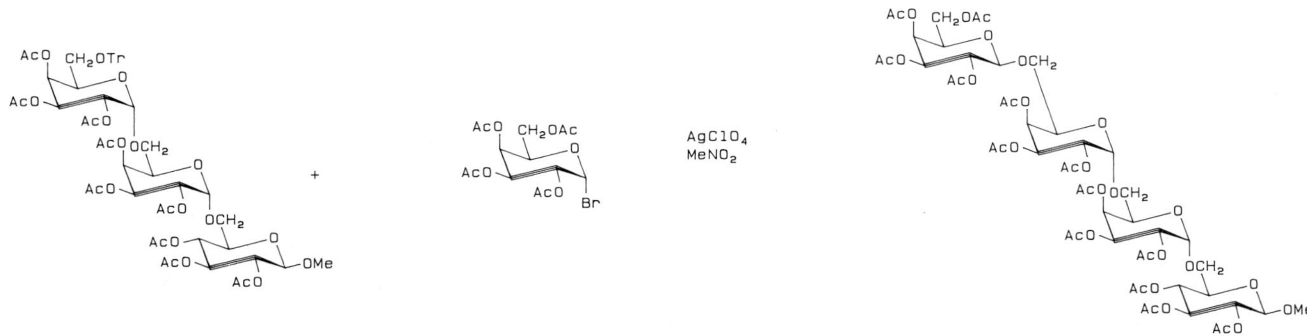

42%; amorphous powder; $[\alpha]_D$ +91.2 (chloroform); Ref.: 30

β-D-Galp-(1→3)-β-D-GlcpNAc-(1→3)-β-D-Galp-(1→4)-D-Glc

m.p. 225–228 °C; $[\alpha]_D$ +27 → +21.3 (water); Ref.: 31, 32

α-acetate; amorphous powder; $[\alpha]_D$ +49.3 (chloroform); Ref.: 31, 32

β-acetate; amorphous powder; $[\alpha]_D$ +22.9 (chloroform); Ref.: 31, 32

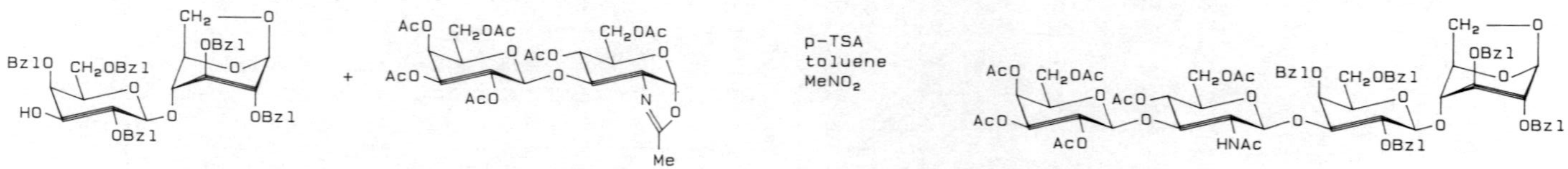

amorphous powder; 77%;
$[\alpha]_D$ −33 (chloroform); Ref.: 31, 32

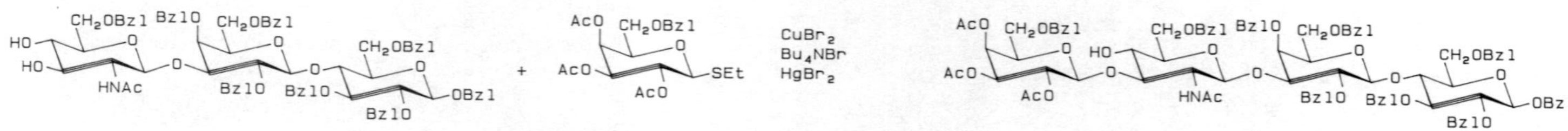

77%; $[\alpha]_D$ −9.2 (chloroform); Ref.: 33

β-D-Galp-(1→4)-β-D-GlcpNAc-(1→3)-β-D-Galp-(1→4)-D-Glc

β-p-(1→OBzl); m.p. 286-288 °C; $[\alpha]_D$ -11 (water); Ref.: 34

β-p-(1→OMe); m.p. 205 °C (dec.); $[\alpha]_D$ +4 (water); Ref.: 37

β-p-(1→OMe); $[\alpha]_D$ +4 (deuteriumoxide); C-13; Ref.: 38

β-p-(1→OMe); C-13; Ref.: 39

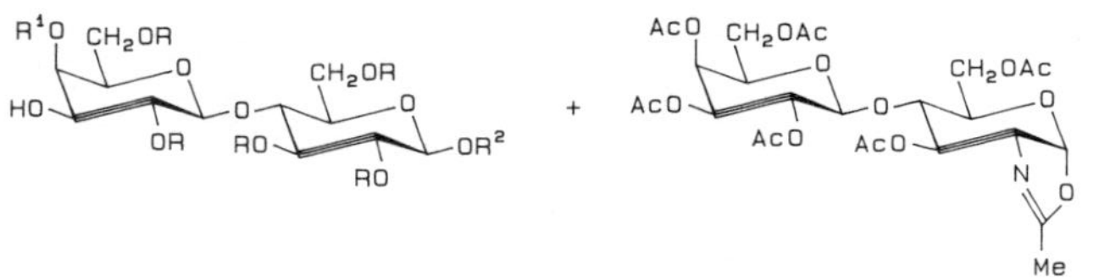

I p-TSA
toluene

II TMSOTf
TMU
dichloro-
ethane

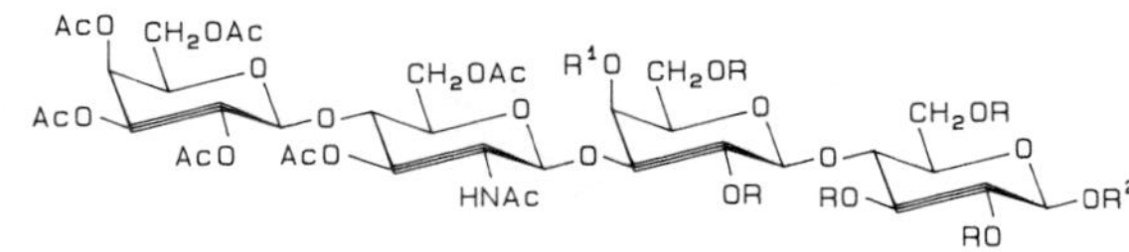

I R=Ac; R^1=H; R^2=Bzl
7% (after deprotection); Ref.: 35

II R=R^1=Bzl; R^2=Me; foam;
$[\alpha]_D$ -9 (chloroform); C-13; Ref.: 38

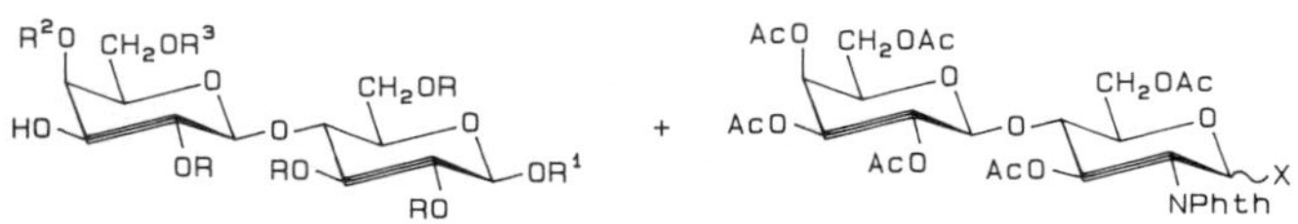

III AgOTf
s-Coll
$MeNO_2$
dichloro-
methane

IV $BF_3.Et_2O$

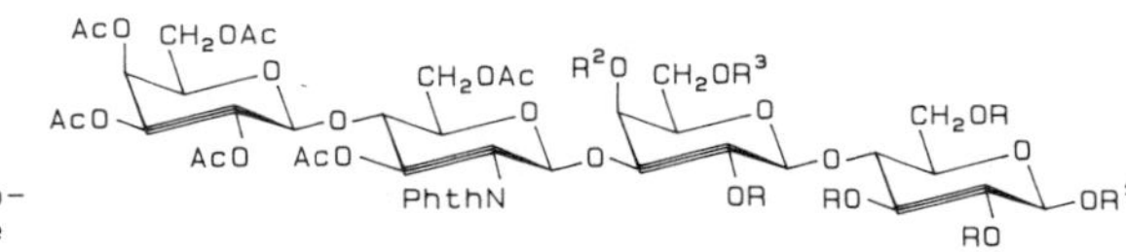

III R^1=Bzl; R=R^2=R^3=Ac; X=β-Cl; 83%
m.p. 126-127 °C; $[\alpha]_D$ -5.3 (chloroform); Ref.: 34

R=R^1=R^3=Bzl; R^2=H; X=Br; 65%
$[\alpha]_D$ +8.6 (chloroform) for its 4'-acetate; Ref.: 36

R^1=Me; R=R^2=R^3=Bzl-p-Br; X=β-Cl; 68%; m.p. 99-102 °C;
$[\alpha]_D$ -37 (chloroform); Ref.: 37

IV R^1=Me; R=Bz; R^2, R^3= CH-Ph; X=β-O-CNH-CCl_3; 75%
m.p. 281-283 °C; $[\alpha]_D$ +37 (chloroform); Ref.: 39

β-D-Galp-(1→4)-β-D-GlcpN-(1→4)-β-D-Galp-(1→4)-D-Glc

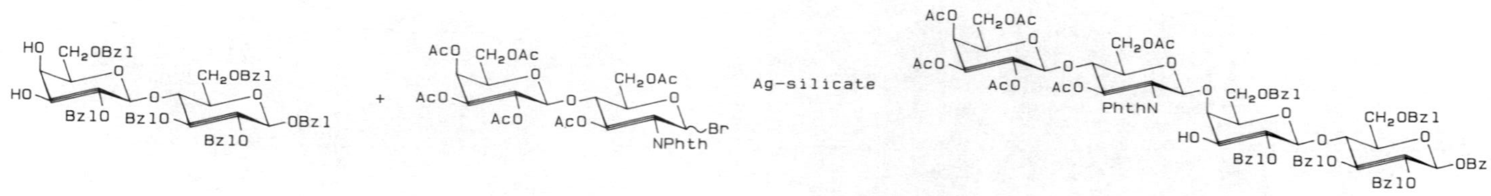

$[\alpha]_D$ +4.7 (chloroform) for its 3'-acetate; Ref.: 36

β-D-Galp-(1→4)-β-D-GlcpNAc-(1→6)-β-D-Galp-(1→4)-D-Glc

m.p. 185-187 °C; $[\alpha]_D$ +8 (water); Ref.: 35

white powder; $[\alpha]_D$ +11.8 (water); Ref.: 40, 41

β-p-(1→OBzl); m.p. 234-236 °C; $[\alpha]_D$ -40 (water); Ref.: 35

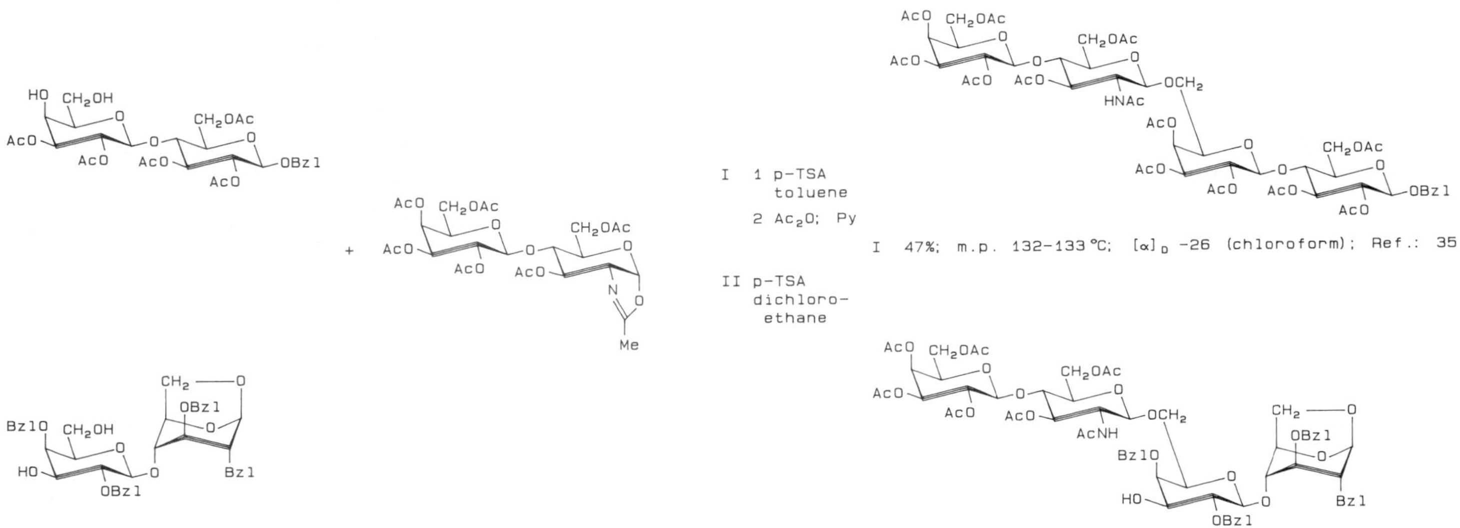

I 47%; m.p. 132-133 °C; $[\alpha]_D$ -26 (chloroform); Ref.: 35

II 24.5%; amorphous powder;
$[\alpha]_D$ -10.8 (chloroform); Ref.: 40, 41

β-D-Galp-(1→4)-α-D-ManpN-(1→3)-β-D-Galp-(1→4)-D-Glc

β-D-Galp-(1→4)-α-D-ManpNAc-(1→3)-β-D-Galp-(1→4)-D-Glc

β-p-(1-OBzl); m.p. 175–177 °C; $[\alpha]_D$ +28 (water); Ref.: 34

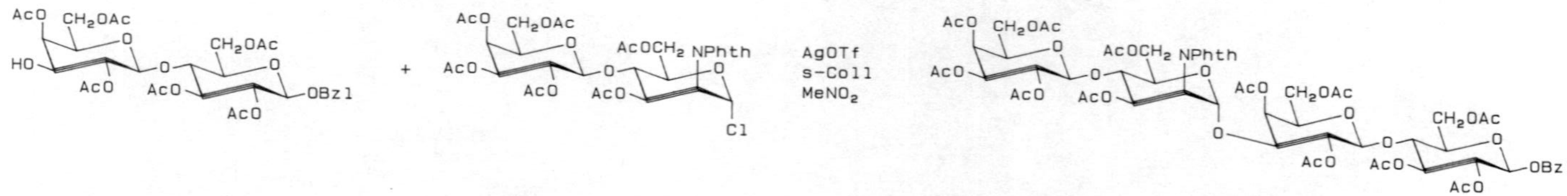

m.p. 121–123 °C; $[\alpha]_D$ −1.3 (chloroform); Ref.: 34

β-D-GalpN-(1→3)-α-D-Galp-(1→4)-β-D-Galp-(1→4)-D-Glc

β-D-GalpNAc-(1→3)-α-D-Galp-(1→4)-β-D-Galp-(1→4)-D-Glc

[α]$_D$ +59.5 (water); Ref.: 42

β-p-(1→OMe); [α]$_D$ +72 (water); C-13; Ref.: 43

β-p-(1→O-octyl); [α]$_D$ +68 (water); C-13; Ref.: 43

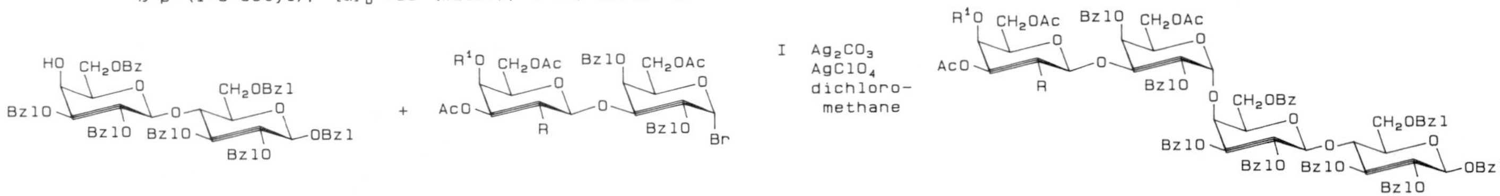

I R=N_3; R[1]=Bzl; 32%
[α]$_D$ +14.8 (chloroform); Ref.: 42

I R=NPhth; R[1]=Ac; 21%
[α]$_D$ -17 (chloroform); Ref.: 42

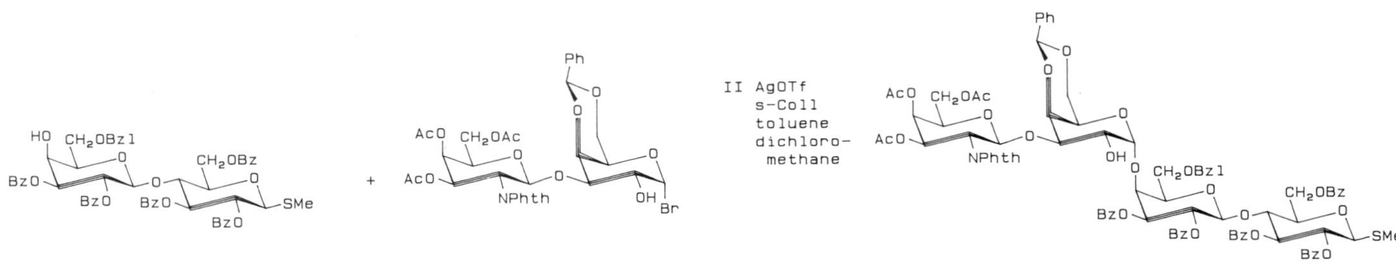

II 66%; [α]$_D$ +49 (chloroform); C-13; Ref.: 43

β-D-Galp-(1→3)-β-D-GalpN-(1→3)-β-D-Galp-(1→4)-D-Glc

β-D-Galp-(1→3)-β-D-GalpNAc-(1→3)-β-D-Galp-(1→4)-D-Glc

$[\alpha]_D$ +39 (MeOH); Ref.: 44

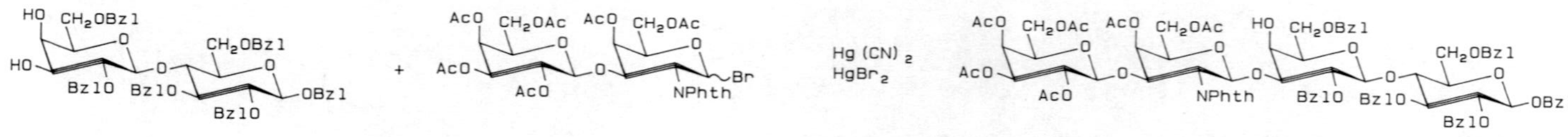

25%; $[\alpha]_D$ +13 (acetone); Ref.: 36, 44

β-D-Galp-(1→3)-β-D-GalpN-(1→4)-β-D-Galp-(1→4)-D-Glc

β-D-Galp-(1→3)-β-D-GalpNAc-(1→4)-β-D-Galp-(1→4)-D-Glc

$[\alpha]_D$ +17 (MeOH); Ref.: 36, 44

$[\alpha]_D$ +21.4 (water); Ref.: 46

β-p-(1→OMCO); powder; $[\alpha]_D$ -9.2 (water); C-13; Ref.: 45

β-p-(1-O 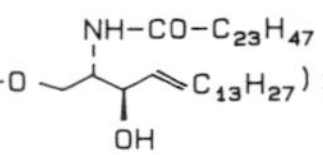); $[\alpha]_D$ +2.5 (chloroform-MeOH 1:1); Ref.: 46

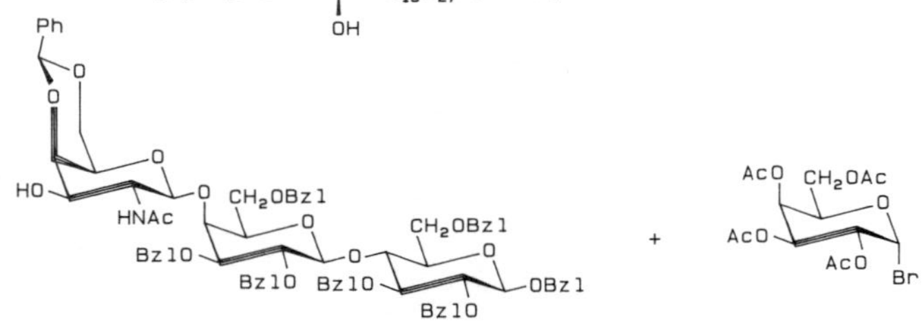

I Hg(CN)$_2$ benzene $MeNO_2$

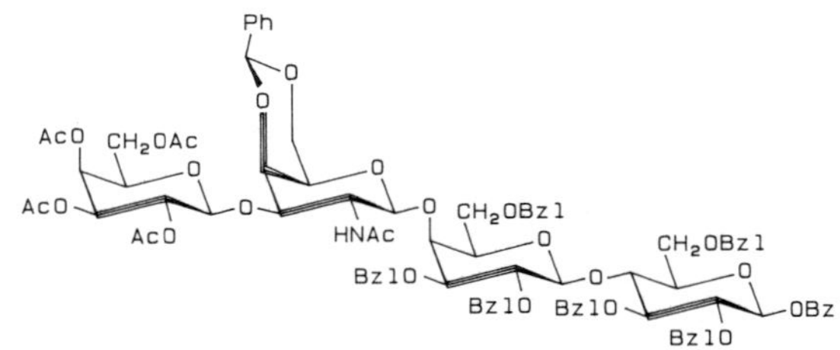

I 97%; $[\alpha]_D$ +47.7 (chloroform); Ref.: 46

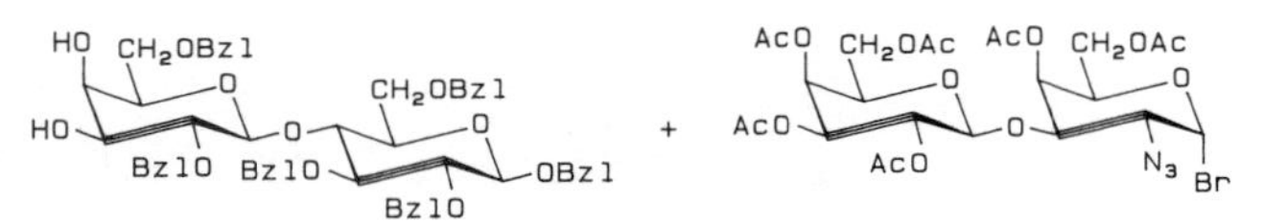

II Ag-silicate or Ag_2CO_3

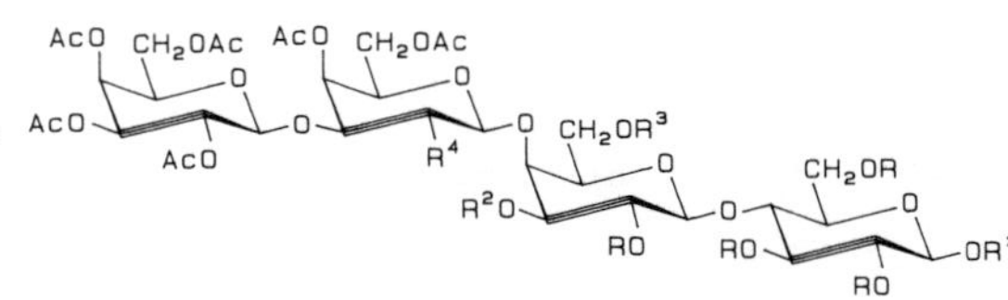

III AgOTf s-Coll $MeNO_2$

II R^1=R=R^3=Bzl; R^2=H; R^4=N_3; 57–62%
m.p. 286 °C; $[\alpha]_D$ +36 (acetone); Ref.: 36, 44

III R^1=MCO; R=R^2=Ac; R^3=Bz; R^4=NPhth; solid
$[\alpha]_D$ -8.3 (chloroform); C-13; Ref.: 45

β-D-GlcpNAc-(1→3)
β-D-GlcpNAc-(1→6) 〉β-D-Galp-(1→4)-D-Glc

m.p. 202–204 °C; $[\alpha]_D$ +13.2 → +15.8 (water); C-13; Ref.: 47

β-p-(1-OMe); $[\alpha]_D$ −13 (water); C-13; Ref.: 39

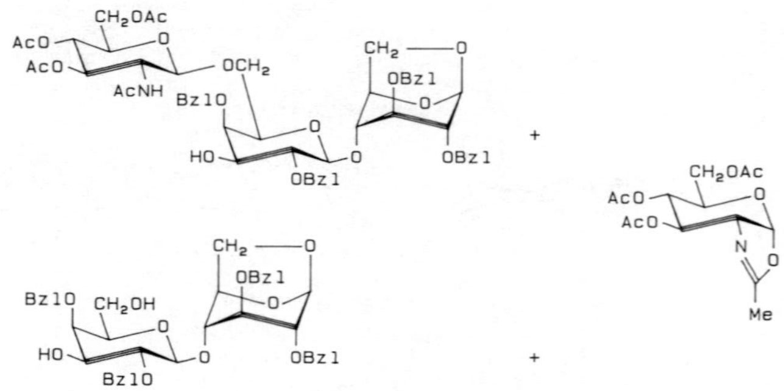

I pTSA toluene $MeNO_2$

II pTSA dichloroethane

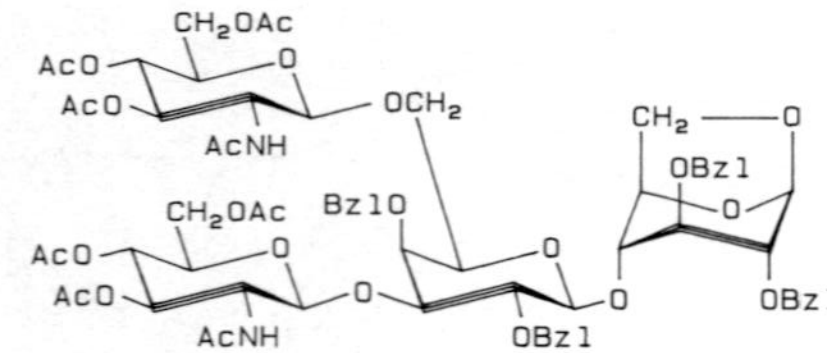

No data; Ref.: 47

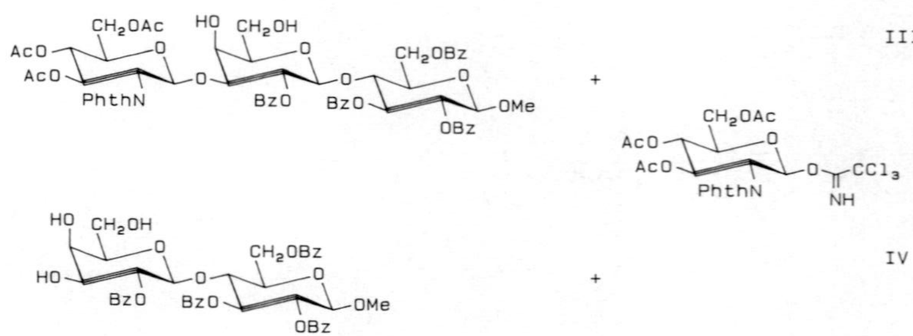

III TMSOTf dichloromethane

IV 1 TMSOTf dichloromethane
2 Ac_2O

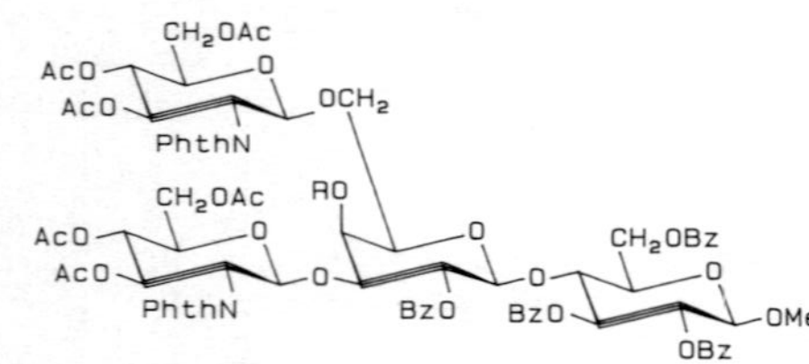

III R=H; 68%; m.p. 162–164 °C; $[\alpha]_D$ +32 (chloroform); Ref.: 39

IV R=Ac; 75%; m.p. 154–155 °C; $[\alpha]_D$ +27 (chloroform); Ref.: 39

β-D-GlcpN-(1→3)
β-D-GalpN-(1→4) 〉β-D-Galp-(1→4)-D-Glc

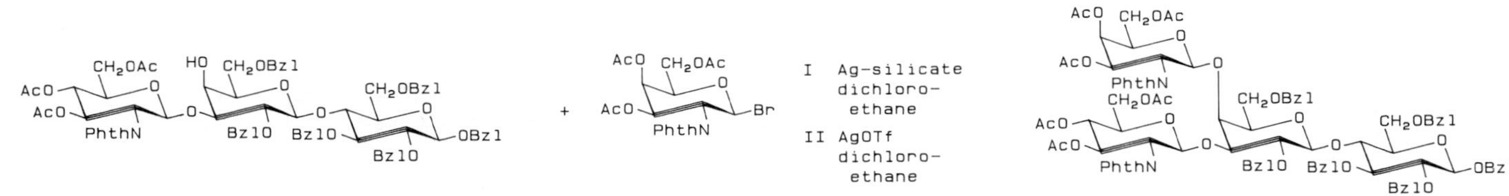

I 76%; $[\alpha]_D$ +32.3 (chloroform); Ref.: 48

II 50%; Ref.: 48

α-Neu5Ac-(2→3)
β-D-GalpN-(1→4) 〉β-D-Galp-(1→4)-D-Glc

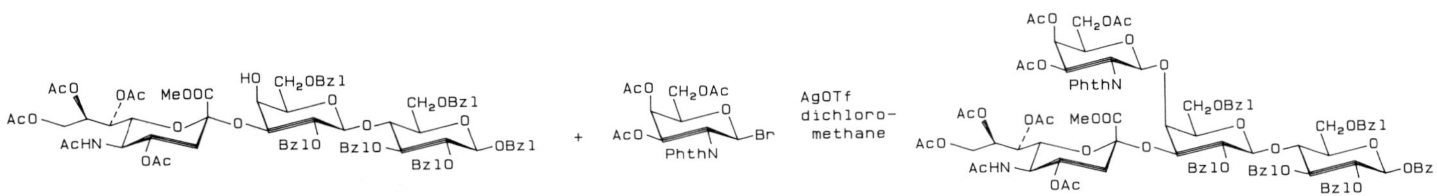

60%; Ref.: 49

β-D-Glcp-(1→4)-α-L-Rhap-(1→3)-β-D-Galp-(1→6)-D-Glc

β-p-(1→OMe); $[\alpha]_D$ −33.5 (water); C-13; Ref.: 50

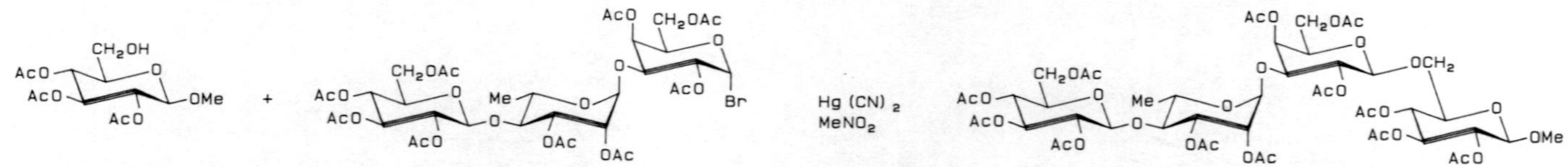

47%; syrup; $[\alpha]_D$ −18.4 (chloroform); Ref.: 50

α-D-GalpNAc-(1→3)
α-D-GlcpNAc-(1→4)
⟩α-D-GalpNAc-(1→4)-D-Glc

amorphous; $[\alpha]_D$ +140 (water); Ref.: 51, 52

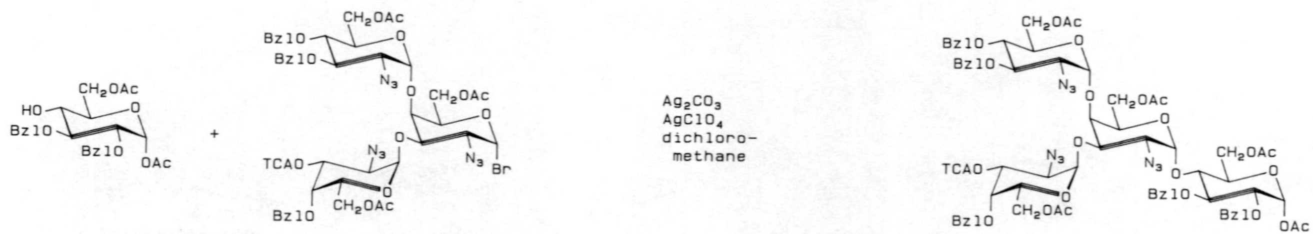

20%; syrup; $[\alpha]_D$ +108 (chloroform); Ref.: 51, 52

β-D-Galp-(1→4)-α-L-Fucp-(1→4)-α-L-Fucp-(1→3)-D-Glc

α-p-(1→OMe); [α]$_D$ -75 (MeOH); C-13; Ref.: 53

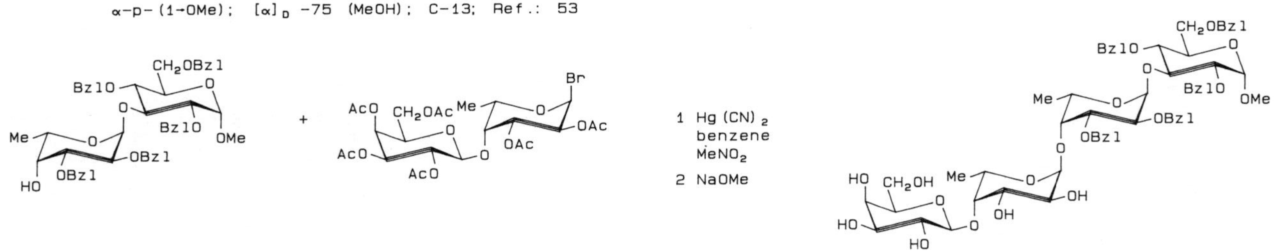

22%; m.p. 178-180 °C; [α]$_D$ -80 (MeOH); Ref.: 53

β-D-Galp-(1→4)-β-L-Fucp-(1→4)-α-L-Fucp-(1→3)-D-Glc

α-p-(1→OMe); [α]$_D$ -23 (MeOH); Ref.: 53

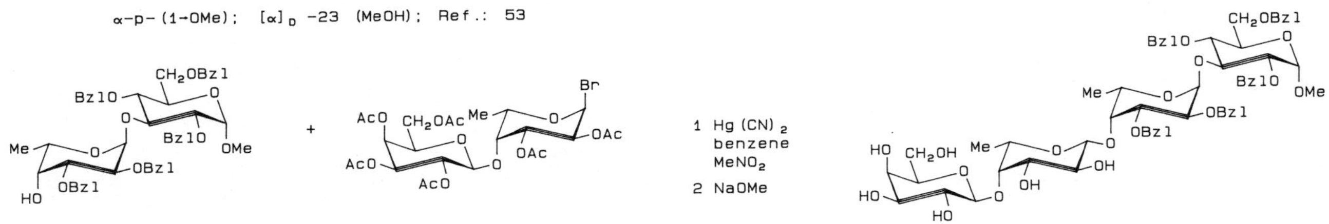

52.5%; m.p. 143-145 °C; [α]$_D$ -40 (MeOH); Ref.: 53

α-L-Fucp-(1→2)-β-D-Galp-(1→4) 〉D-Glc
α-L-Fucp-(1→3)

amorphous powder; $[\alpha]_D$ −103.2 (water); C-13; Ref.: 54

white solid; $[\alpha]_D$ −100 (water); C-13; Ref.: 55

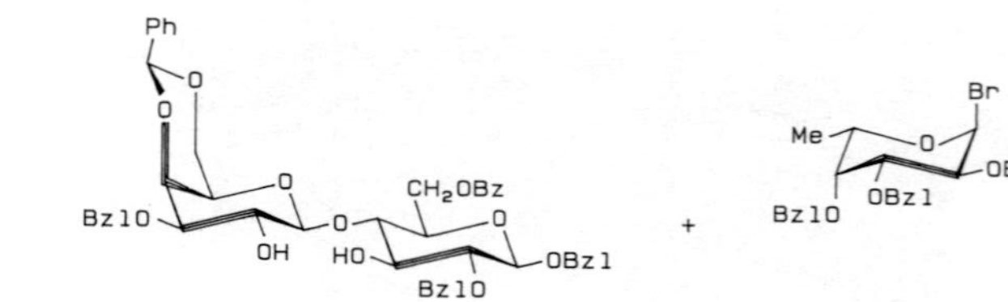

I 1 Et_4NBr
DMF
dichloroethane
2 NaOMe

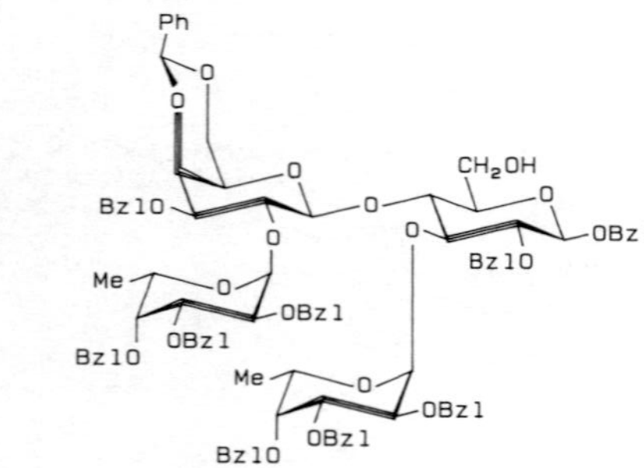

I amorphous powder; 68%
$[\alpha]_D$ −86.9 (chloroform); C-13; Ref.: 54

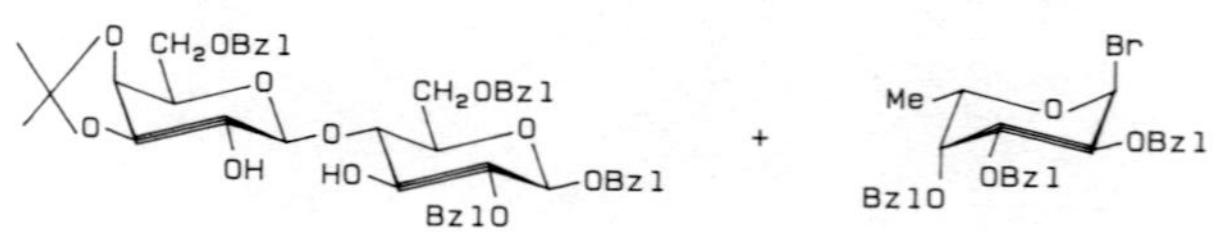

II Et_4NBr
DMF
dichloromethane

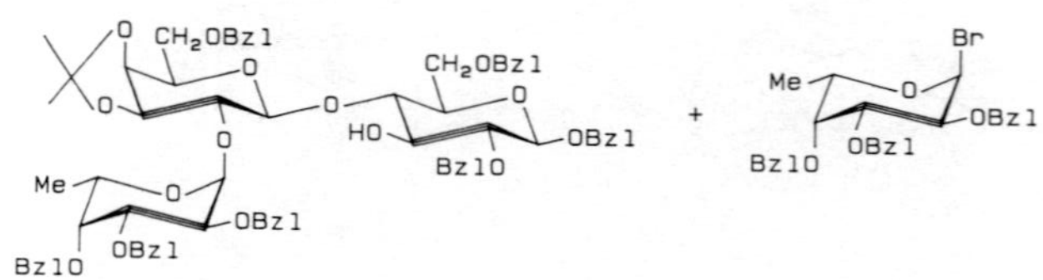

III $HgBr_2$
dichloromethane
Me(iPr)$_2$Py

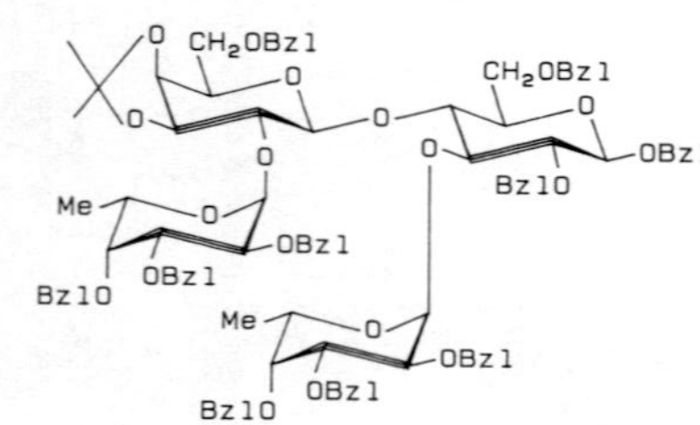

II 34%; $[\alpha]_D$ −59.5 (chloroform); C-13; Ref.: 55

III 73%; Ref.: 55

β-D-Glcp-(1→2)
α-L-Rhap-(1→4) → D-Glc
β-D-Glcp-(1→6)

m.p. 179–180 °C; $[\alpha]_D$ −16 (water); C-13; Ref.: 56

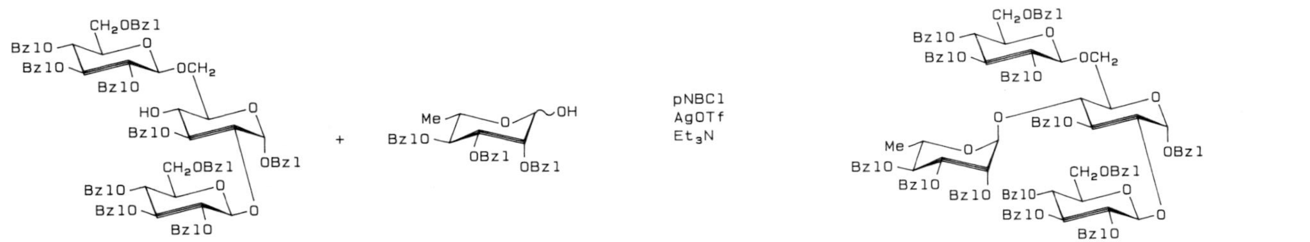

37%; $[\alpha]_D$ +19 (chloroform); C-13; Ref.: 56

α-D-GlcpN-(1→6)-α-D-GlcpN-(1→6)-α-D-GlcpN-(1→6)-D-GlcN

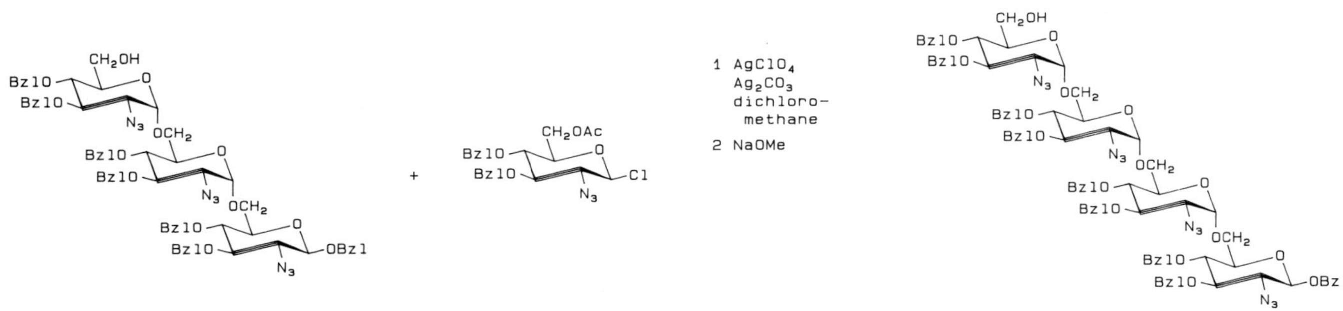

39%; $[\alpha]_D$ +78 (chloroform); Ref.: 57

β-D-GlcpNAc-(1→4)-β-D-GlcpNAc-(1→6)-β-D-GlcpNAc-(1→4)-D-GlcNAc

β-p-(1→OPNP); m.p. 248–249 °C; $[\alpha]_D$ −28 (water); Ref.: 58

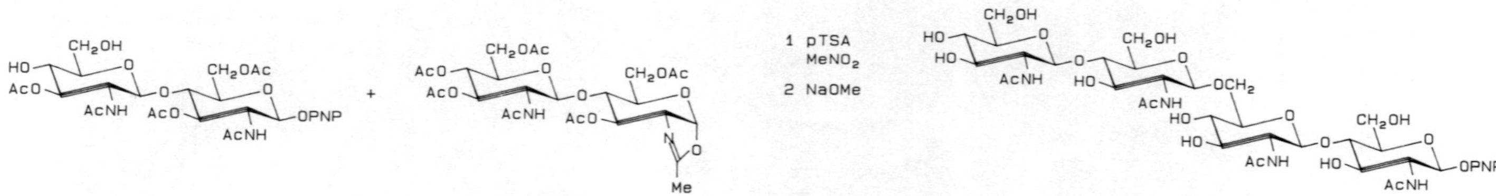

13%; Ref.: 58

α-D-Manp-(1→3)-β-D-Manp-(1→4)-β-D-GlcpNAc-(1→4)-D-GlcNAc

α-Phosphate acetate; m.p. 157–159 °C; $[\alpha]_D$ 0.0 (Chloroform-MeOH 5:1); Ref.: 59

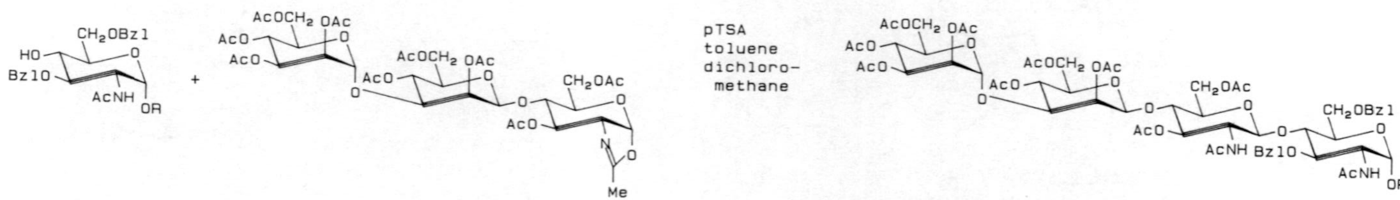

R=All; 22%; m.p. 95–97 °C;
$[\alpha]_D$ +5 (chloroform-MeOH 5:1); Ref.: 59, 61

R=Bzl; 32%; Ref.: 60

R=Bzl; 19%; m.p. 79–83 °C;
$[\alpha]_D$ +13 (dichloroethane); Ref.: 59

α-KDO-(2→4)-α-KDO-(2→6)-β-D-GlcpN-(1→6)-D-GlcN

$[\alpha]_D$ +61 (water); C-13; Ref.: 62

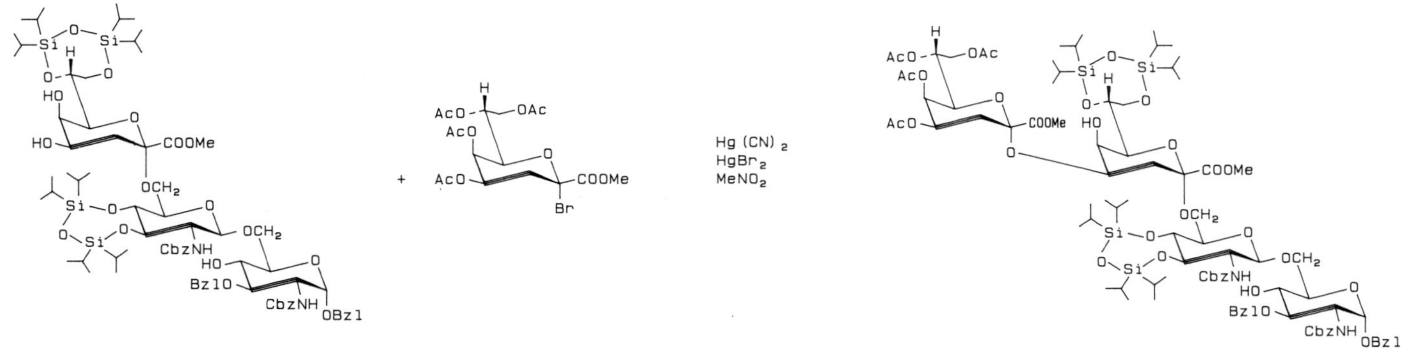

30%; Ref.: 62

α-D-Manp-(1→2)-α-D-Manp-(1→3)-β-D-Manp-(1→4)-D-GlcNAc

acetate; white powder; $[\alpha]_D$ +18 (chloroform); C-13; Ref.: 63

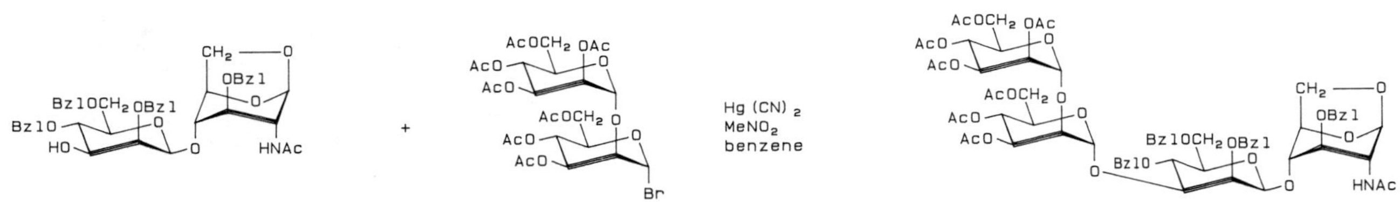

40%; foamy solid; $[\alpha]_D$ -21.8 (chloroform); C-13; Ref.: 63

α-D-Manp-(1→3)-α-D-Manp-(1→6)-β-D-Manp-(1→4)-D-GlcNAc

acetate; white powder; $[\alpha]_D$ +51 (chloroform); C-13; Ref.: 63

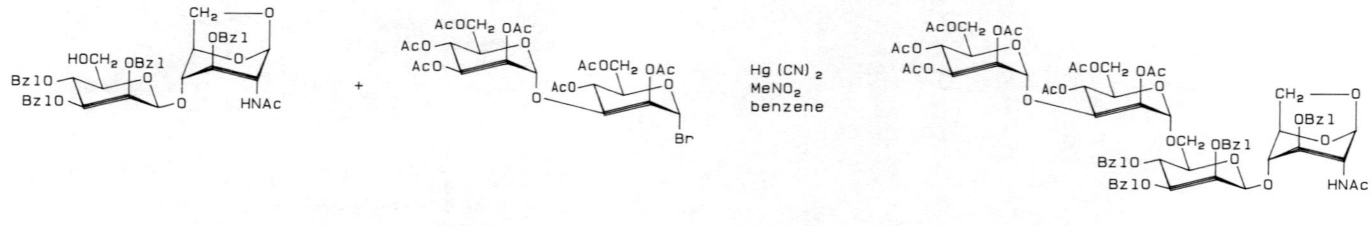

29.4%; foamy solid; $[\alpha]_D$ −13.3 (chloroform); Ref.: 63

α-D-Manp-(1→3)
α-D-Manp-(1→6) 〉β-D-Manp-(1→4)-D-GlcNAc

syrup; $[\alpha]_D$ +37.2 (MeOH); Ref.: 64, 65

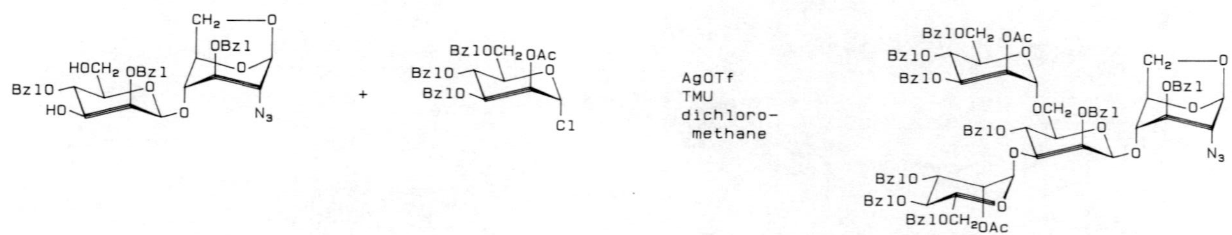

80%; $[\alpha]_D$ +14.2 (chloroform); Ref.: 64

82%; syrup;
$[\alpha]_D$ +12.9 (chloroform); C-13; Ref.: 65

α-L-Rhap-(1→2)-α-L-Rhap-(1→3)-α-L-Rhap-(1→3)-D-GlcNAc

β-p-(1→OMCO); $[\alpha]_{589}$ −58.1 (water); C-13; Ref.: 66

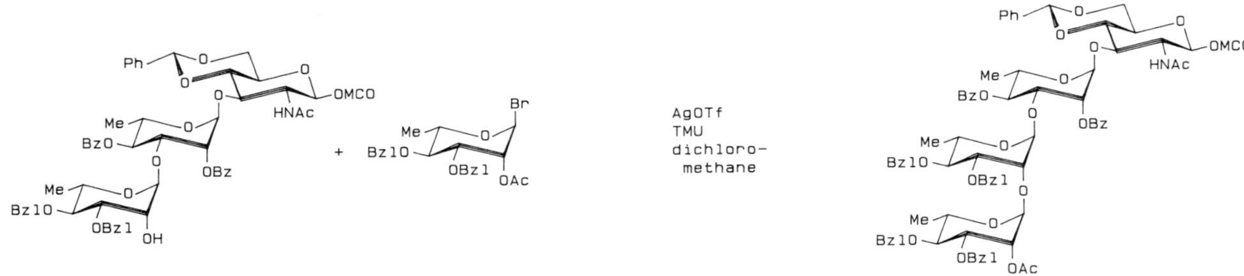

70%; $[\alpha]_{589}$ +8.1 (chloroform); C-13; Ref.: 66

β-D-GlcpA-(1→4)-α-D-GlcpN-(1→4)-α-L-IdopA-(1→4)-D-GlcN

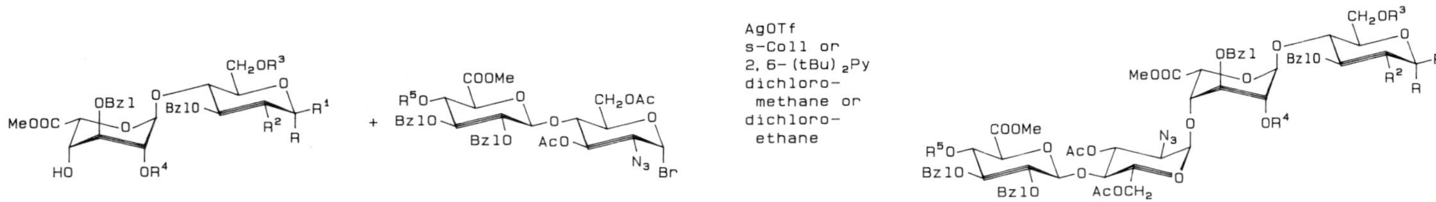

R=H; R^1=OBzl; R^2=N_3; R^3=Ac; R^4=Ac; R^5=Lev; 68%; Ref.: 70

R=H; R^1=OBzl; R^2=N_3; R^3=Ac; R^4=Bz; R^5=MCA; 37%; Ref.: 71

R=OBzl; R^1=H; R^2=NHCbz; R^3=Ac; R^4=Ac; R^5=MCA;
55%; $[\alpha]_D$ +57 (chloroform); Ref.: 69

R=OBzl; R^1=H; R^2=NHCbz; R^3=Ac; R^4=Ac; R^5=MCA;
after dechloroacetylation (R^5=H); 30%;
$[\alpha]_D$ +62 (chloroform); Ref.: 68

R=OMe; R^1=H; R^2=NHCbz; R^3=Bzl; R^4=Ac; R^5=Lev; 47%; Ref.: 67

α-L-Fucp-(1→2)
α-D-Galp-(1→3) 〉β-D-Galp-(1→3)-D-GlcNAc

amorphous; $[\alpha]_D$ +33.3 (MeOH); Ref.: 72, 73

$[\alpha]_D$ +35 (MeOH); Ref.: 74

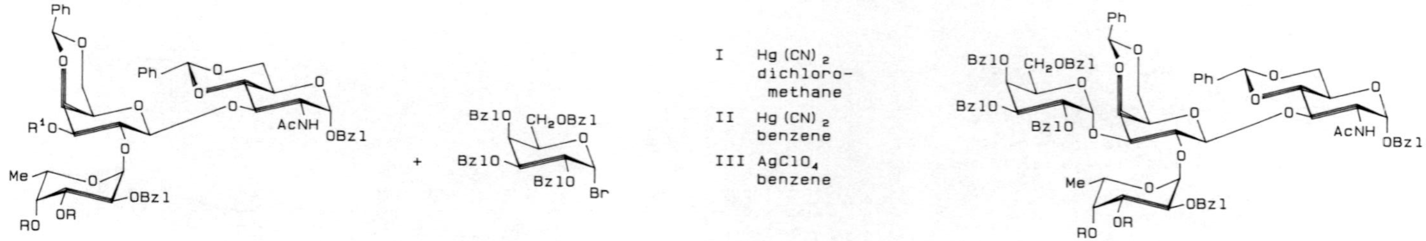

I R=Bzl; R^1=H; 78%; m.p. 97 °C; $[\alpha]_D$ +20.2 (chloroform); Ref.: 72, 73

II R=NBz; R^1=H; 66%; m.p. 10 °C; $[\alpha]_D$ −7 (chloroform); Ref.: 74

III R=NBz; R^1=Dpc; 60%; Ref.: 74

α-L-Fucp-(1→2)
α-D-Galp-(1→4) 〉β-D-Galp-(1→3)-D-GlcNAc

β-p-(1→OPr); amorphous solid; $[\alpha]_D$ −2.0 (MeOH); C-13; Ref.: 75

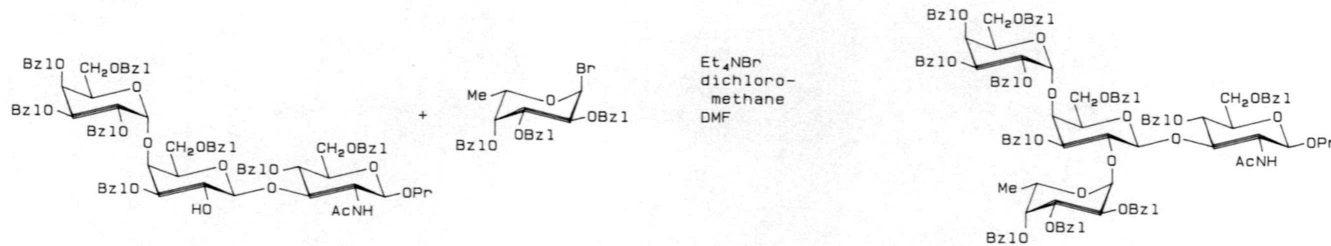

68%; glass; $[\alpha]_D$ +3.4 (chloroform); $[\alpha]_{436}$ +7.7 (chloroform); Ref.: 75

α-L-Fucp-(1→2)
α-D-GalpNAc-(1→3) ⟩ β-D-Galp-(1→3)-D-GlcNAc

amorphous; $[\alpha]_D$ +53.8 (MeOH); Ref.: 72, 73

amorphous; $[\alpha]_D$ +49 (MeOH); Ref.: 76, 77

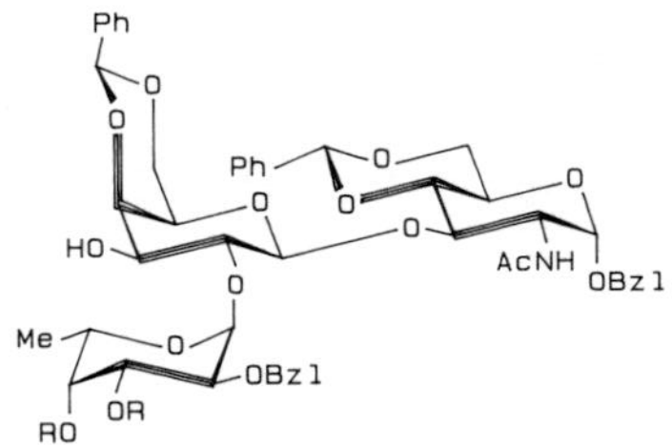

+

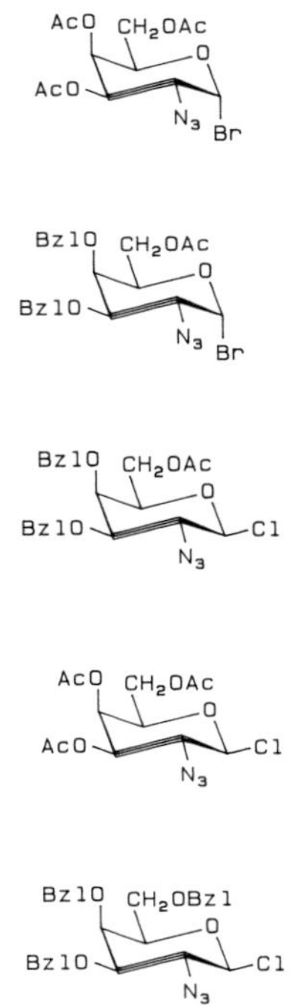

I 1 $Hg(CN)_2$, $HgBr_2$, dichloromethane; 2 NaOMe, MeOH

II $Hg(CN)_2$, dichloromethane

III Ag_2CO_3, $AgClO_4$, dichloromethane

IV 1 Ag_2CO_3, $AgClO_4$, dichloromethane; 2 NaOMe, MeOH

V Ag_2CO_3, $AgClO_4$, dichloromethane

VI AgOTf, Ag_2CO_3, dichloromethane

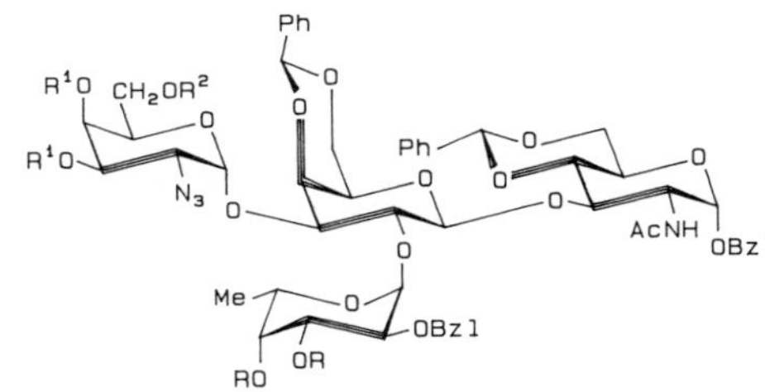

I R=Bzl; R^1=R^2=H; 63%; m.p. 128 °C; $[\alpha]_D$ +79.8 (chloroform); Ref.: 72, 73

II R=R^1=Bzl; R^2=Ac; 64%; m.p. 92 °C; $[\alpha]_D$ +66.8 (chloroform); Ref.: 73

III R=R^1=Bzl; R^2=Ac; 53%; syrup; Ref.: 73

IV R=Bzl; R^1=R^2=H; 46%; m.p. 128 °C; $[\alpha]_D$ +79.8 (chloroform); Ref.: 73

V R=R^1=R^2=Bzl; 60%; syrup; $[\alpha]_D$ +26 (chloroform); Ref.: 76, 77

R=NBz; R^1=R^2=Bzl; 82%; m.p. 151–153 °C; $[\alpha]_D$ +9 (chloroform); Ref.: 76, 77, 78

VI R=NBz; R^1=R^2=Bzl; 92%; m.p. 151–153 °C; $[\alpha]_D$ +10 (chloroform); Ref.: 74

β-D-Galp-(1→4)-β-D-GlcpNAc-(1→3)-β-D-Galp-(1→4)-D-GlcNAc

m.p. 195–198 °C (dec.); $[\alpha]_D$ +21 → +14 (water); Ref.: 79

β-p-(1→OMe); m.p. 259–261 °C; $[\alpha]_D$ −10 (water); Ref.: 80

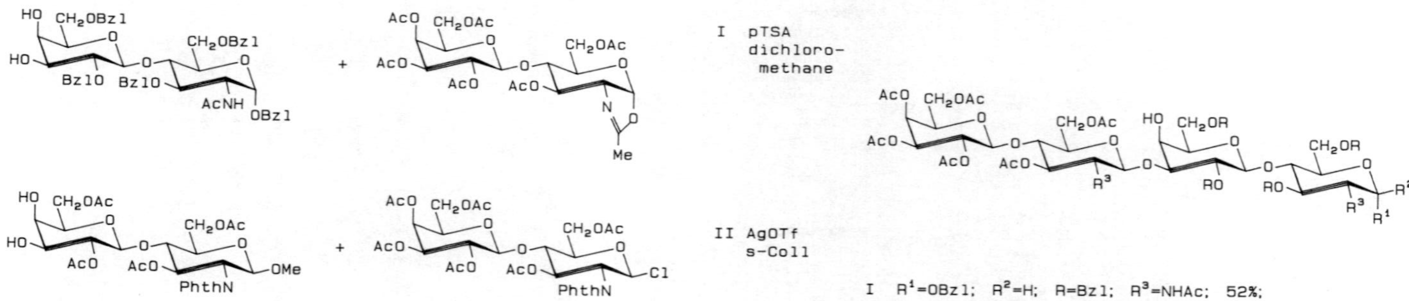

I R^1=OBzl; R^2=H; R=Bzl; R^3=NHAc; 52%;
m.p. 215–216 °C; $[\alpha]_D$ +39 (chloroform); Ref.: 79

II R^1=H; R^2=OMe; R=Ac; R^3=NPhth; 64%;
m.p. 203–205 °C; $[\alpha]_D$ +20.5 (chloroform); Ref.: 80

α-L-Fucp-(1→2)
α-D-Galp-(1→3) 〉β-D-Galp-(1→4)-D-GlcNAc

amorphous; $[\alpha]_D$ +14.5 (water); Ref.: 83, 84

$[\alpha]_D$ +14.1 (water); Ref.: 81, 82

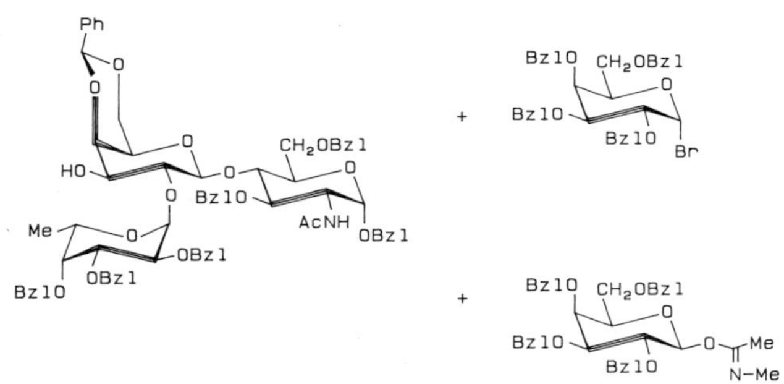

I AgOTf, Ag_2CO_3, dichloromethane

II pTSA, $MeNO_2$

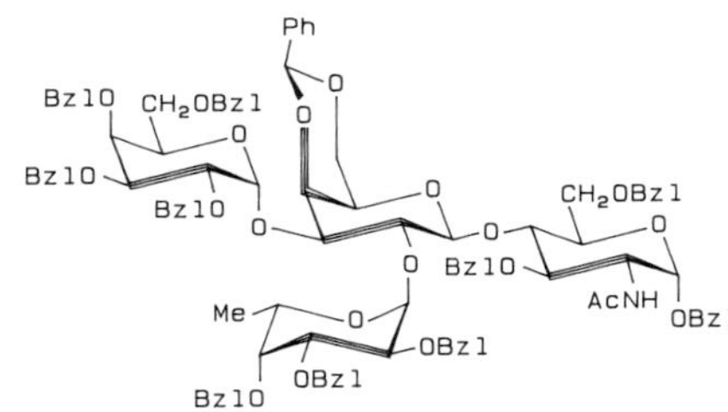

I 82%; $[\alpha]_D$ +27.8 (chloroform); Ref.: 83

I 82%; syrup; $[\alpha]_D$ +32.3 (chloroform); Ref.: 84

II 90%; $[\alpha]_D$ +34 (chloroform); Ref.: 81, 82

α-L-Fucp-(1→2)
α-D-Galp-(1→4) 〉β-D-Galp-(1→4)-D-GlcNAc

β-p-(1→OPr); $[\alpha]_D$ −19.1 (MeOH); Ref.: 85

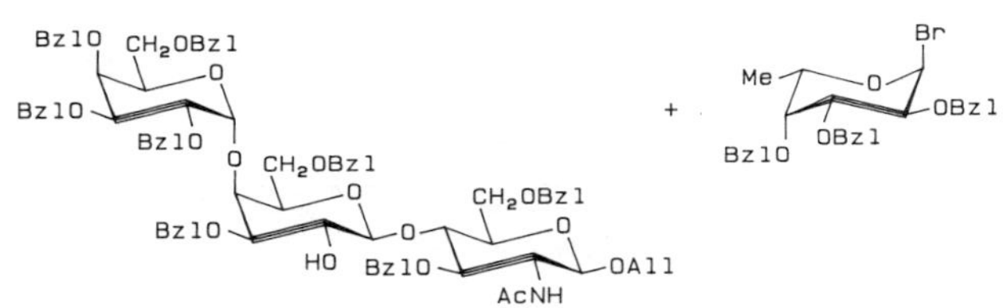

Et_4NBr, dichloromethane, DMF

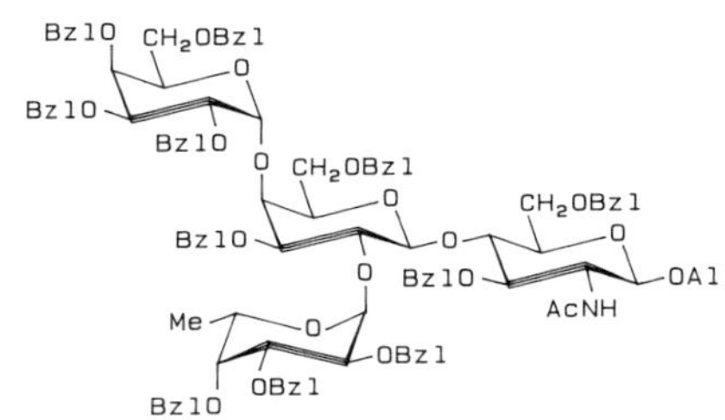

58%; $[\alpha]_D$ −2.8 (chloroform); Ref.: 85

α-L-Fucp-(1→2) ⟩ β-D-Galp-(1→4)-D-GlcNAc
α-D-GalpNAc-(1→3) ⟩

$[\alpha]_D$ +21.3 (water); Ref.: 83

amorphous; $[\alpha]_D$ +45 (water); Ref.: 84

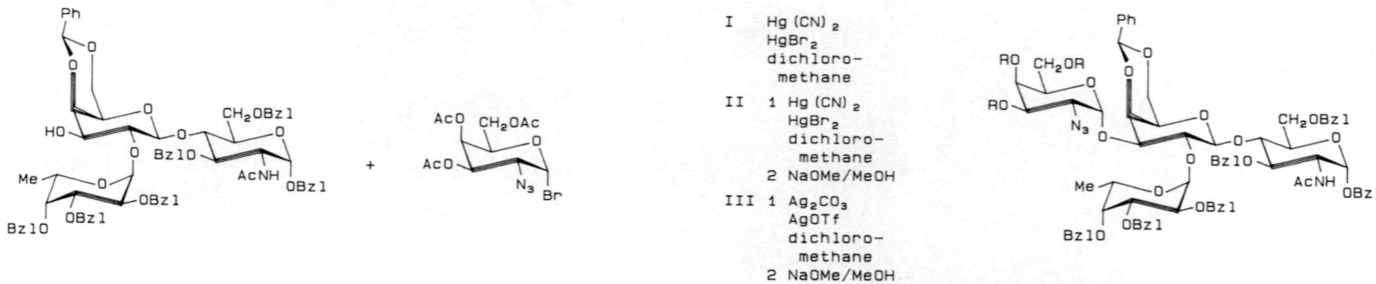

I R=Ac; 88%; $[\alpha]_D$ +47.6 (chloroform); Ref.: 83

II R=H; 65%; amorphous;
$[\alpha]_D$ +29.6 (chloroform); Ref.: 84

III R=H; 67%; amorphous; Ref.: 84

β-D-Galp-(1→4)-β-D-Glcp-(1→6)
β-D-Galp-(1→4) 〉D-GlcNAc

β-O⁀O⁀O⁀COOMe; $[\alpha]_D$ −12 (MeOH-water); Ref.: 86

β-O⁀O⁀O⁀NHAc; $[\alpha]_D$ −13 (MeOH); Ref.: 86

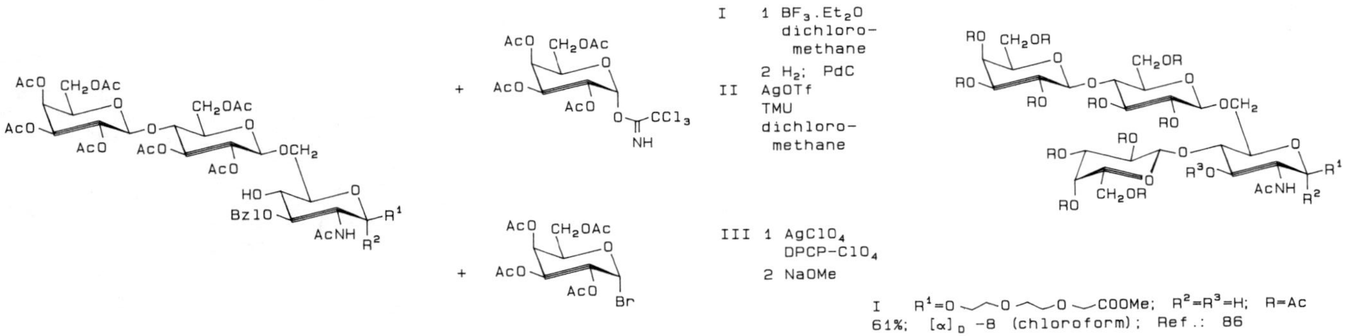

I R^1=O⁀O⁀O⁀COOMe; R^2=R^3=H; R=Ac
61%; $[\alpha]_D$ −8 (chloroform); Ref.: 86

II R^1=O⁀O⁀O⁀N_3; R^2=H; R^3=Bzl; R=Ac
31%; m.p. 94–95 °C; $[\alpha]_D$ −13.5 (chloroform); Ref.: 86

III R^1=R=H; R^2=OBzl; R^3=Bzl;
49%; $[\alpha]_D$ +58 (MeOH); Ref.: 87

α-L-Fucp-(1→2)-β-D-Galp-(1→3)
α-L-Fucp-(1→4) 〉D-GlcNAc

amorphous; $[\alpha]_D$ -75.7 (water); C-13; Ref.: 88

$[\alpha]_D$ -69 (water); Ref.: 74

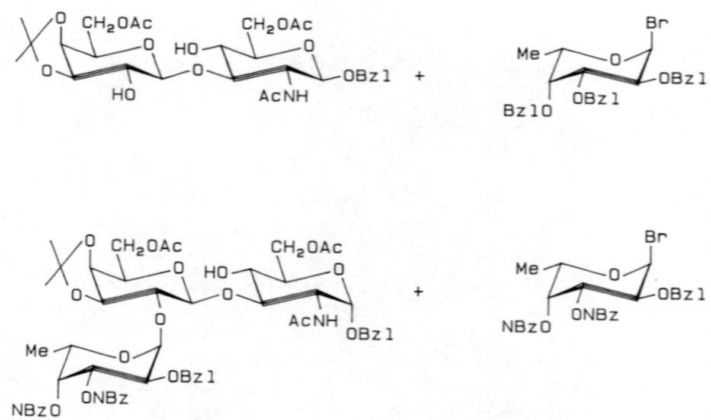

Et$_4$NBr
dichloromethane
DMF

Hg(CN)$_2$
MeNO$_2$
benzene

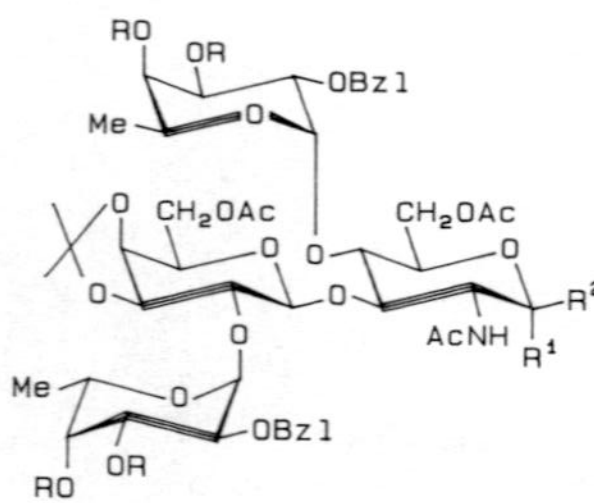

R^1=H; R^2=OBzl; R=Bzl; 78%; amorphous
$[\alpha]_D$ -60.4 (chloroform); C-13; Ref.: 88

R^1=OBzl; R^2=H; R=NBz; 74%; amorphous
$[\alpha]_D$ -201 (chloroform); Ref.: 74

α-L-Fucp-(1→2)-β-D-Galp-(1→4) \
α-L-Fucp-(1→3) / D-GlcNAc

m.p. 214-216 °C (dec.); $[\alpha]_D$ -113 → -124.5 (MeOH-water 1:9); Ref.: 90

β-p-(1→OMCO); $[\alpha]_D$ -104 (water); C-13; Ref.: 89

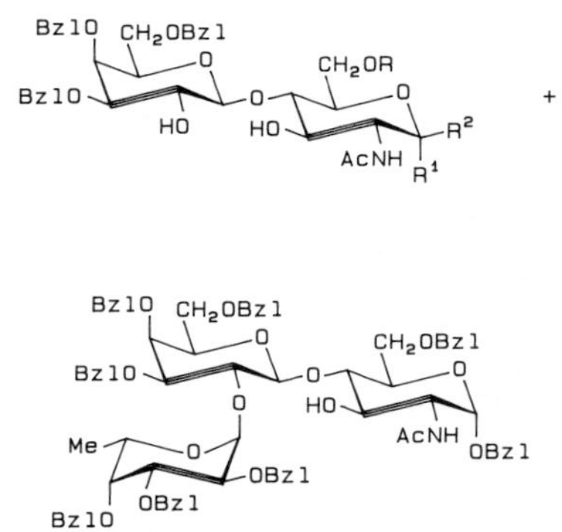

+

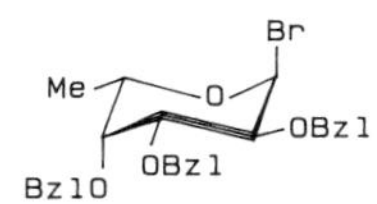

I Et$_4$NBr, dichloromethane, DMF

II Et$_4$NBr, dichloroethane, DMF, (iPr)$_2$EtN

III Et$_4$NBr, dichloroethane, DMF, (iPr)$_2$EtN

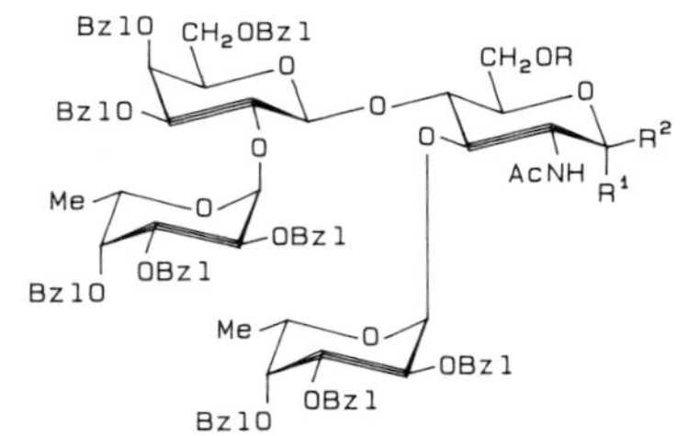

I R^1=H; R^2=OMCO; R=CH$_2$OBzl;
75%; $[\alpha]_D$ -66 (chloroform); Ref.: 89

II R^1=OBzl; R^2=H; R=Bzl;
74%; colourless glass;
$[\alpha]_D$ -31.5 (chloroform); Ref.: 90

III R^1=OBzl; R^2=H; R=Bzl;
80%; Ref.: 90

α-D-Glcp-(1→2)-α-D-Glcp-(1→3)-α-D-Glcp-(1→3)-D-Man

[α]$_D$ +37.5 (water); Ref.: 91

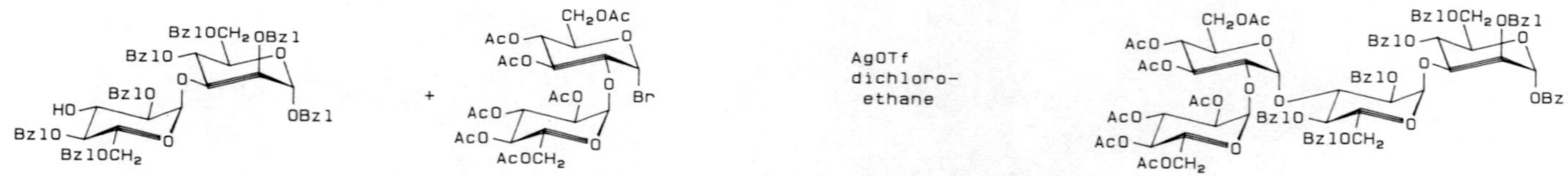

AgOTf
dichloro-
ethane

59%; [α]$_D$ +104.4 (chloroform); C-13; Ref.: 91

α-D-Neu5Ac-(2→6)-β-D-Galp-(1→4)-β-D-GlcpNAc-(1→2)-D-Man

[α]$_D$ -20 (water); C-13; Ref.: 92

amorphous; [α]$_D$ -21 (water); Ref.: 93, 94

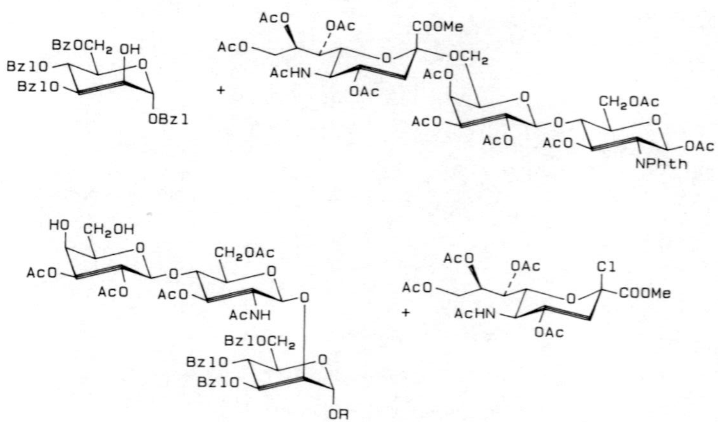

I TMSOTf
dichloro-
methane

II Hg(CN)$_2$
HgBr$_2$
dichloro-
ethane

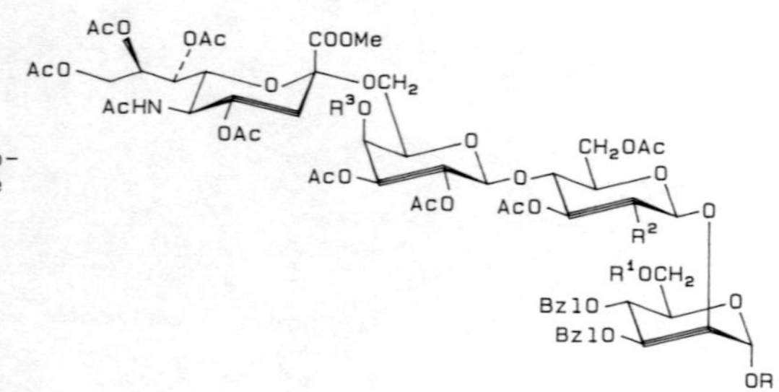

I R=Bzl; R^1=Bz; R^2=NPhth; R^3=Ac
47%; [α]$_D$ +16.3 (chloroform); Ref.: 93, 94

II R=R^1=Bzl; R^2=NHAc; R^3=H
34%; [α]$_D$ +0.5 (chloroform); Ref.: 92

R=All; R^1=Bzl; R^2=NHAc; R^3=H
48%; [α]$_D$ -2.6 (chloroform); Ref.: 95

β-D-Neu5Ac-(2→6)-β-D-Galp-(1→4)-β-D-GlcpNAc-(1→2)-D-Man

[α]$_D$ -21.4 (water); C-13; Ref.: 92

amorphous; [α]$_D$ -25 (water); Ref.: 93, 94

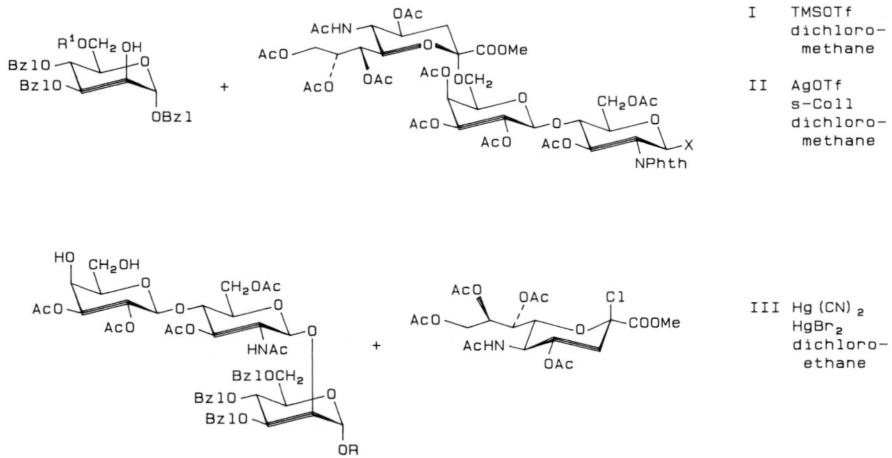

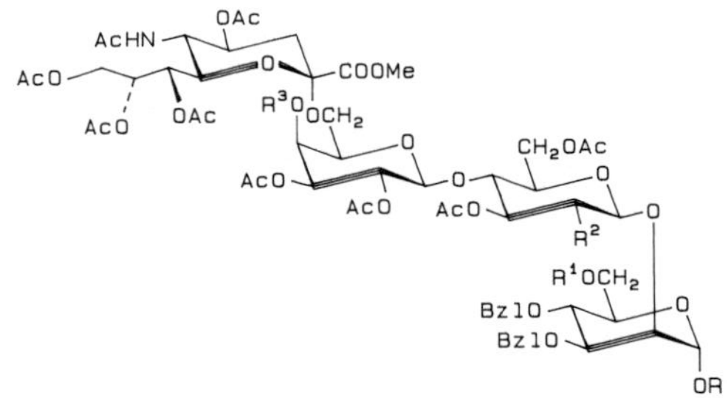

I R=Bzl; R^1=R^3=Ac; R^2=NPhth; X=OAc
76%; [α]$_D$ -18.3 (chloroform); Ref.: 93, 94

II R=R^1=Bzl; R^2=NPhth; R^3=Ac
X=Cl 15%; X=Br 22%
[α]$_D$ +7.7 (chloroform); Ref.: 93

III R=R^1=Bzl; R^2=NHAc; R^3=H
30%; [α]$_D$ +4 (chloroform); Ref.: 92

III R=All; R^1=Bzl; R^2=NHAc; R^3=H
33%; [α]$_D$ +2.1 (chloroform); Ref.: 95

α-L-Fucp-(1→3)
β-D-Galp-(1→4) 〉β-D-GlcpNAc-(1→2)-D-Man

amorphous powder; $[\alpha]_{578}$ −78 (water); C-13; Ref.: 96

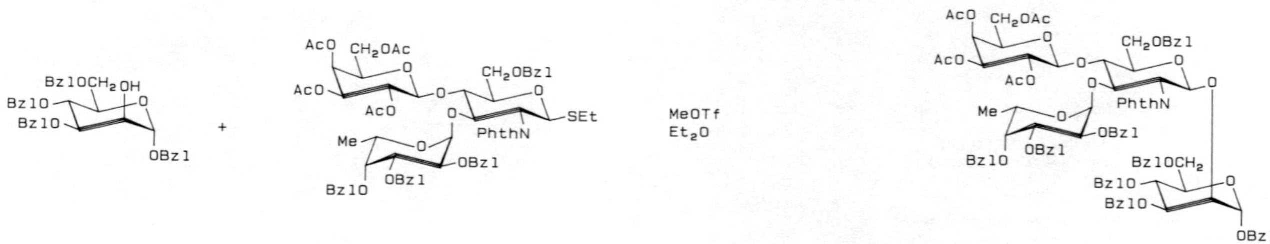

67%; $[\alpha]_{578}$ −9 (chloroform); C-13; Ref.: 96

α-D-Manp-(1→2)-α-D-Manp-(1→2)-α-D-Manp-(1→2)-D-Man

$[\alpha]_D$ +31.3 (water); C-13; Ref.: 97

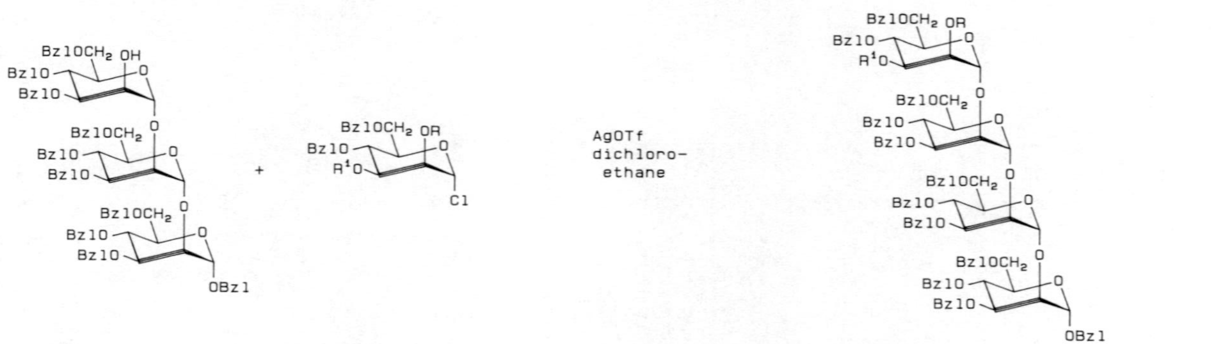

R=Ac; R^1=Bz1
89%; $[\alpha]_D$ +23 (chloroform); Ref.: 97

R=Bz1; R^1=Ac
80%; $[\alpha]_D$ +20 (chloroform); Ref.: 98

α-D-Manp-(1→3)-α-D-Manp-(1→3)-α-D-Manp-(1→3)-D-Man

amorphous solid; $[\alpha]_D$ +25.4 (water); C-13; Ref.: 99

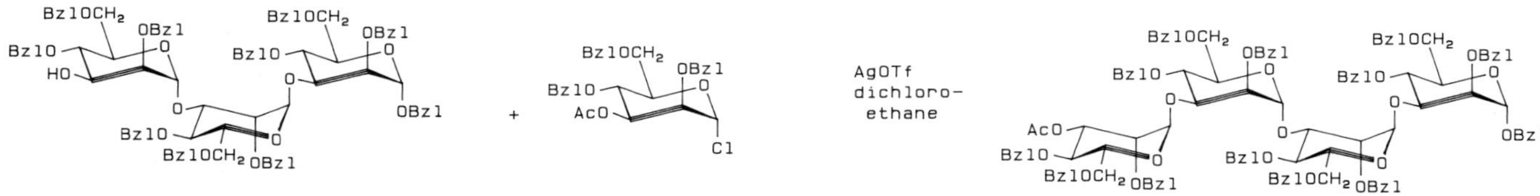

88%; $[\alpha]_D$ +22.4 (chloroform); Ref.: 99

α-D-Manp-(1→2)-α-D-Manp-(1→6)-α-D-Manp-(1→6)-D-Man

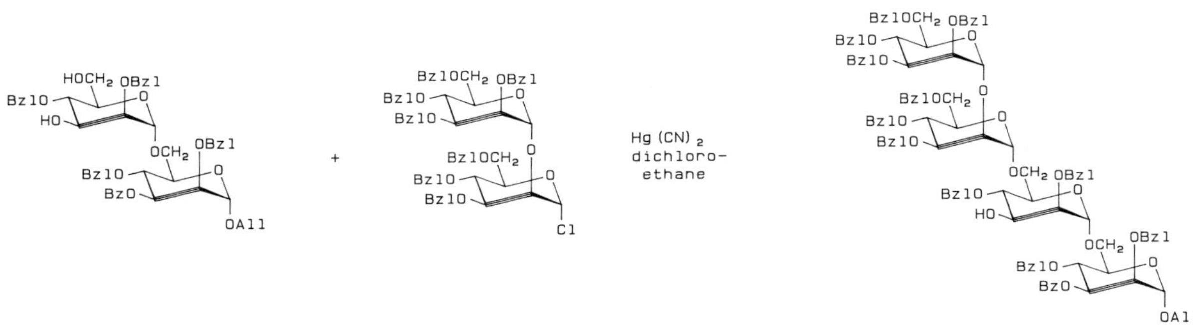

22%; $[\alpha]_D$ +37 (chloroform); C-13; Ref.: 100

α-D-Manp-(1→2)-β-D-Manp-(1→6)-α-D-Manp-(1→6)-D-Man

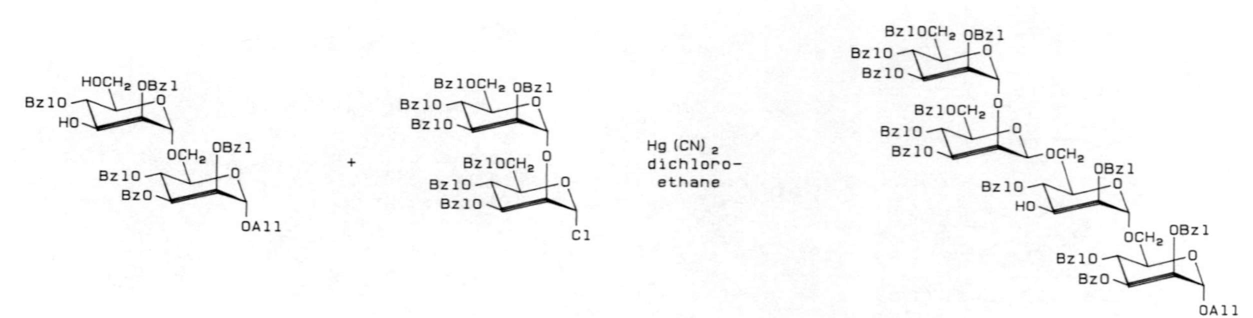

11%; $[\alpha]_D$ +6.3 (chloroform); C-13; Ref.: 100

β-D-GlcpNAc-(1→2)
β-D-GlcpNAc-(1→6)
α-D-Manp-(1→6)-D-Man

β-p-(1→OCD₃); $[\alpha]_D$ -30.74 (MeOH); Ref.: 101

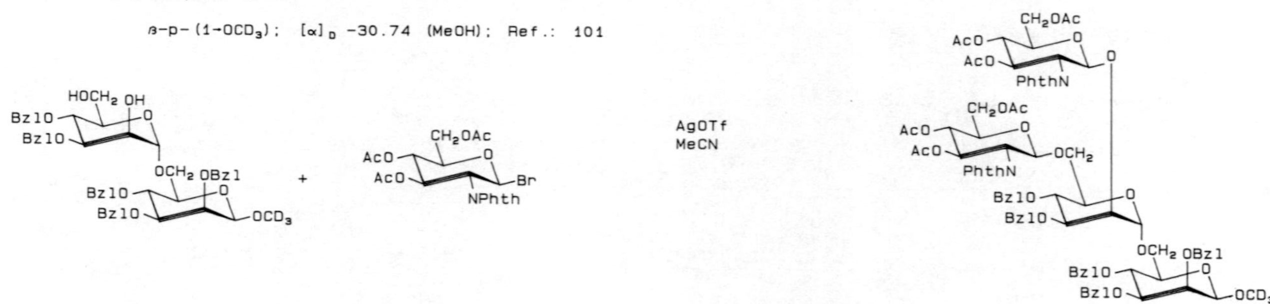

26%; $[\alpha]_D$ -10.3 (chloroform); Ref.: 101

α-D-Manp-(1→3)
α-D-Manp-(1→6) ⟩ α-D-Manp-(1→6)-D-Man

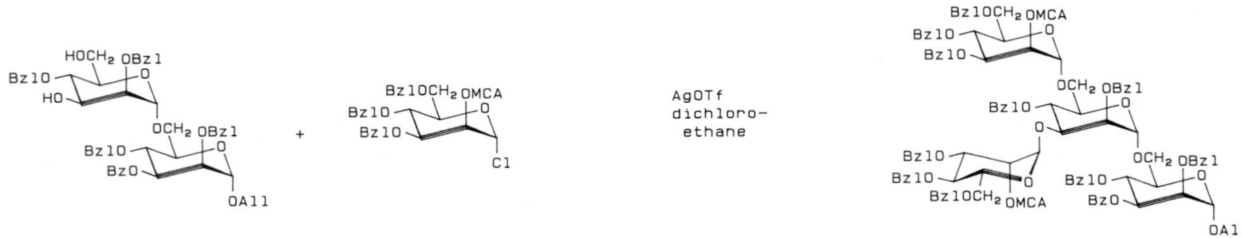

68.6%; $[\alpha]_D$ +10 (chloroform); C-13; Ref.: 100

α-D-Manp-(1→3)
β-D-GlcpNAc-(1→2)-α-D-Manp-(1→6) ⟩ D-Man

β-p-(1→OMCO); $[\alpha]_D$ +6.48 (water); C-13; Ref.: 102

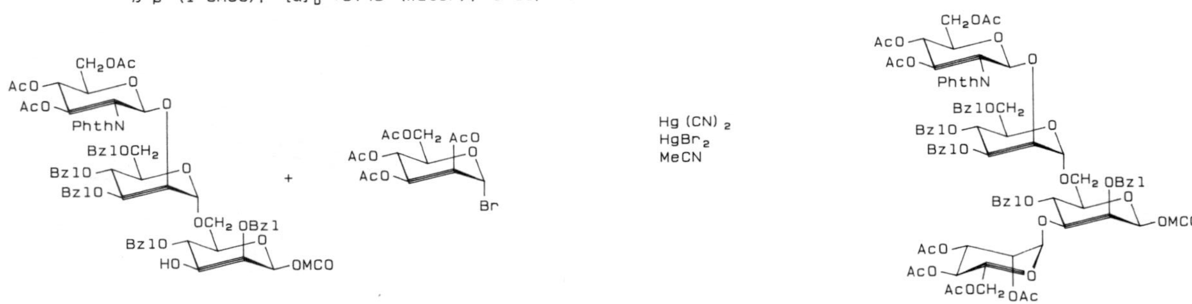

65%; $[\alpha]_D$ +4.41 (chloroform); C-13; Ref.: 102

β-D-GlcpNAc-(1→2)-α-D-Manp-(1→6) (1→3) ? 〉D-Man
α-D-Manp-(1→3) (1→6) ?

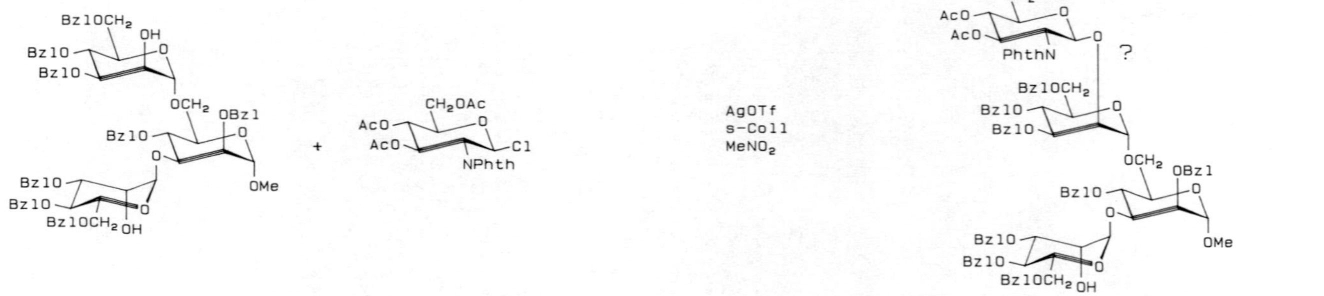

27%; $[\alpha]_D$ +25.1 (chloroform); Ref.: 103

α-D-Manp-(1→3)-α-D-Manp-(1→6) 〉D-Man
α-D-Manp-(1→3)

α-p-(1→OMe); $[\alpha]_D$ +71.3 (water); Ref.: 104

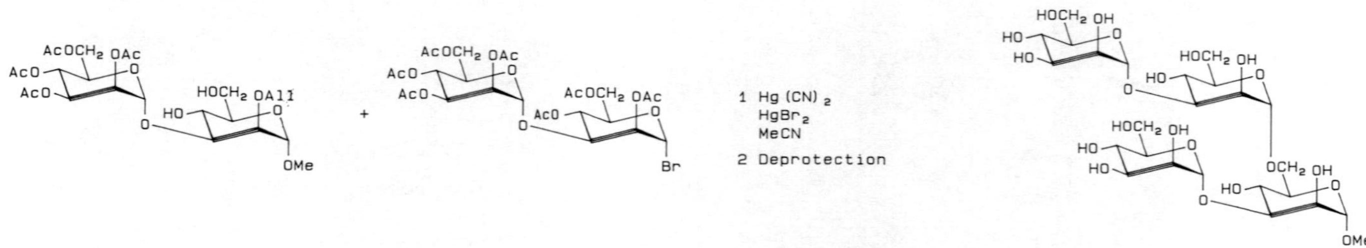

16%; Ref.: 104

α-D-Manp-(1→3)
β-D-GlcpNAc-(1→4) →D-Man
α-D-Manp-(1→6)

amorphous powder; $[\alpha]_D$ +55 (water); C-13; Ref.: 105

58%; syrup; $[\alpha]_D$ +53 (chloroform); C-13; Ref.: 105

α-D-Manp-(1→2)
α-D-Manp-(1→4) →D-Man
α-D-Manp-(1→6)

white powder; $[\alpha]_D$ +77.4 (water); C-13; Ref.: 106

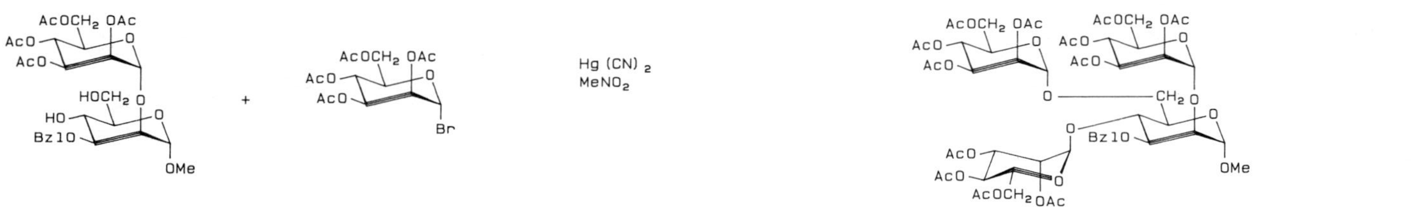

85%; m.p. 75-76 °C; $[\alpha]_D$ +34.1 (chloroform); C-13; Ref.: 106

α-D-Glcp-(1→4)-α-D-Glcp-(1→3)-α-L-Rhap-(1→3)-D-ManNAc

β-p-(1→OMCO); $[\alpha]_D$ +43.6 (MeOH); Ref.: 107

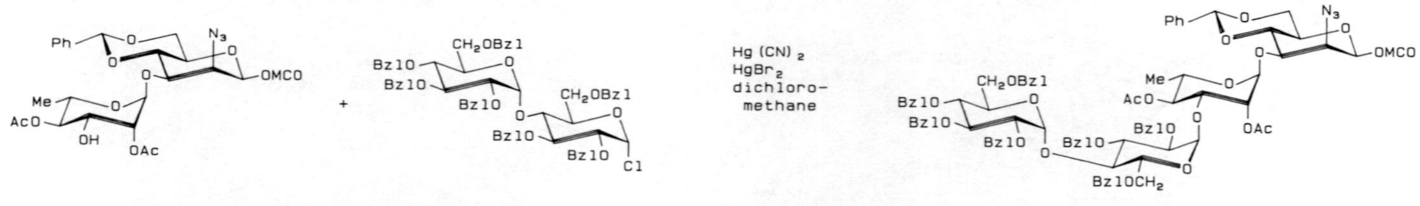

48%; Ref.: 107

α-L-Rhap-(1→3)-α-L-Rhap-(1→3)-β-D-GlcpNAc-(1→2)-L-Rha

α-p-(1→OMCO); $[\alpha]_D$ -55.3 (MeOH); C-13; Ref.: 108

α-p-[1→O(CH$_2$)$_8$CONH-NH$_2$]; $[\alpha]_D$ -53.1 (MeOH); Ref.: 108

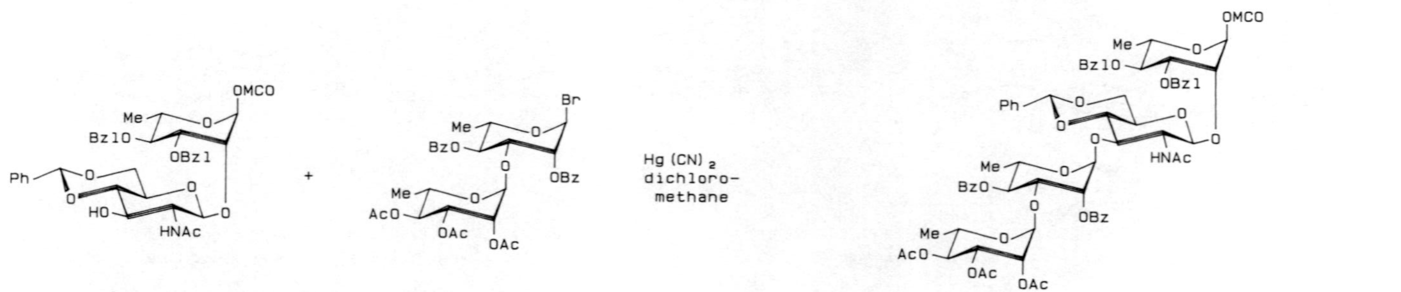

90.6%; syrup; $[\alpha]_D$ -8.7 (dichloromethane); C-13; Ref.: 108

α-D-Galp-(1→2) \
α-D-Parp-(1→3) / α-D-Manp-(1→4)-L-Rha

α-p-(1→OMCO); $[\alpha]_D$ +80 (water); C-13; Ref.: 109

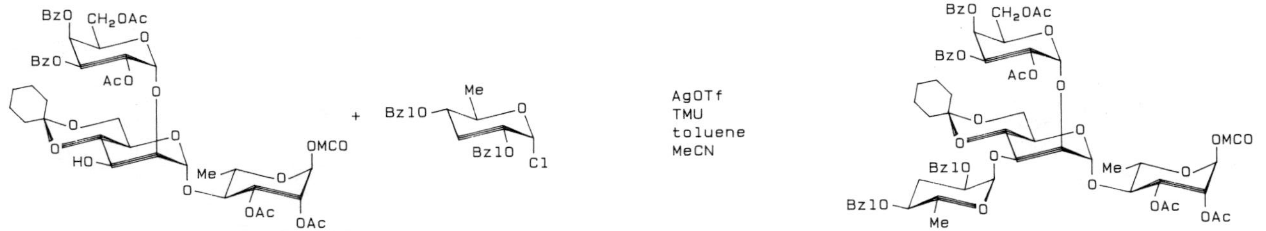

Parp=3,6-dideoxy-D-ribo-hexopyranosyl

36%; $[\alpha]_D$ +100 (chloroform); Ref.: 109

α-D-Galp-(1→2) \
β-D-Parp-(1→3) / α-D-Manp-(1→4)-L-Rha

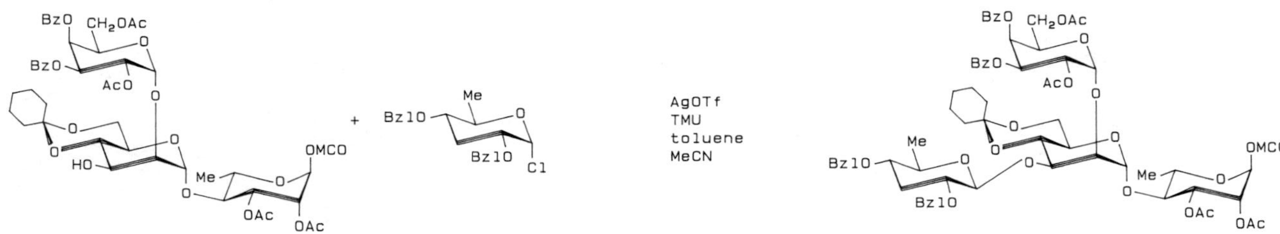

Parp=3,6-dideoxy-D-ribo-hexopyranosyl

Ref.: 109

α-D-Galp-(1→2)
α-D-Tyvp-(1→3) 〉α-D-Manp-(1→4)-L-Rha

[α]$_D$ +75 (water); C-13; Ref.: 109

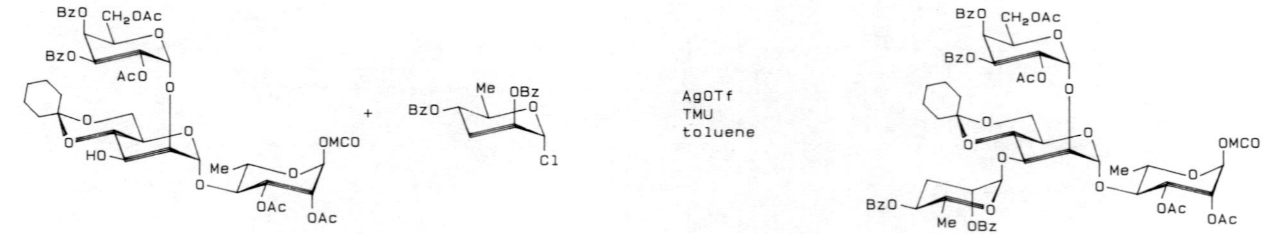

Tyvp=3,6-dideoxy-D-arabino-hexopyranosyl

94%; [α]$_D$ +44 (chloroform); Ref.: 109

α-D-Galp-(1→2)
α-D-Abep-(1→3) 〉α-D-Manp-(1→4)-L-Rha

α-p-(1→OMCO); [α]$_D$ +64 (water); C-13; Ref.: 110

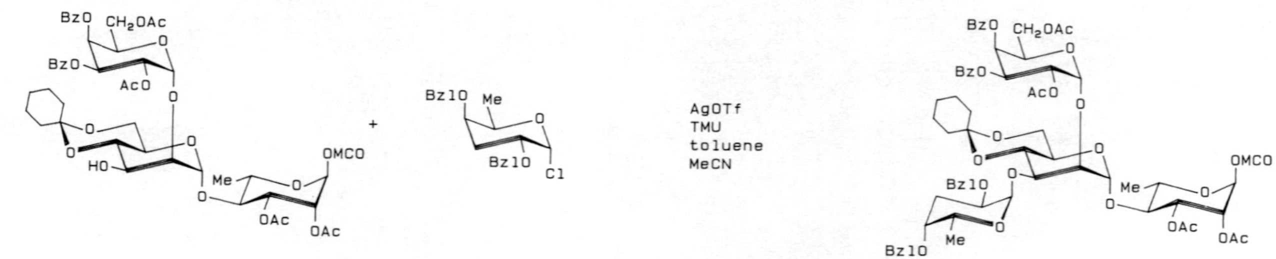

D-Abep=3,6-dideoxy-D-xylo-hexopyranosyl

62%; [α]$_D$ +82 (chloroform); C-13; Ref.: 110

α-D-Galp-(1→2) \
β-D-Abep-(1→3) / α-D-Manp-(1→4)-L-Rha

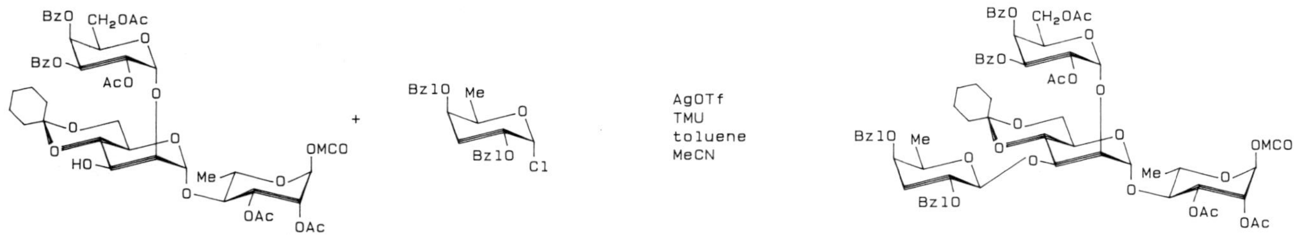

Abep=3,6-dideoxy-D-xylo-hexopyranosyl

14%; C-13; Ref.: 110

α-D-Galp-(1→2) \
α-L-Ascp-(1→3) / α-D-Manp-(1→4)-L-Rha

α-p-(1→OMCO); $[\alpha]_D$ +9 (water); C-13; Ref.: 109

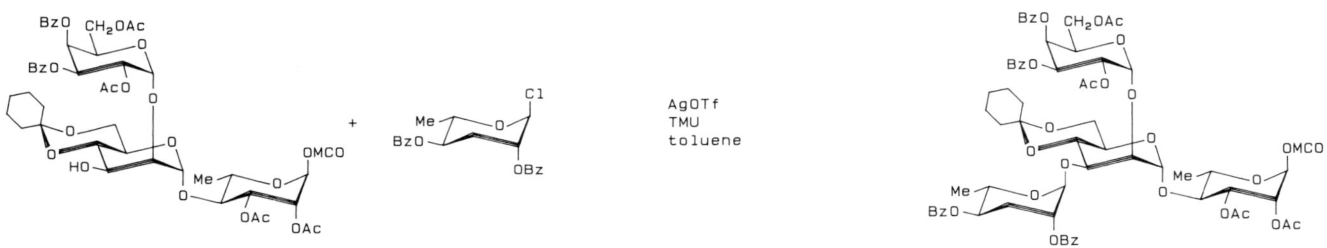

L-Ascp=3,6-dideoxy-L-arabino-hexopyranosyl

100%; $[\alpha]_D$ +54 (chloroform); Ref.: 109

α-D-Galp-(1→2)
2,3,6-trideoxy-α-D-threo-Hexp-(1→3) 〉α-D-Manp-(1→4)-L-Rha

α-p-(1→OMCO); $[\alpha]_D$ +62 (water); C-13; Ref.: 109

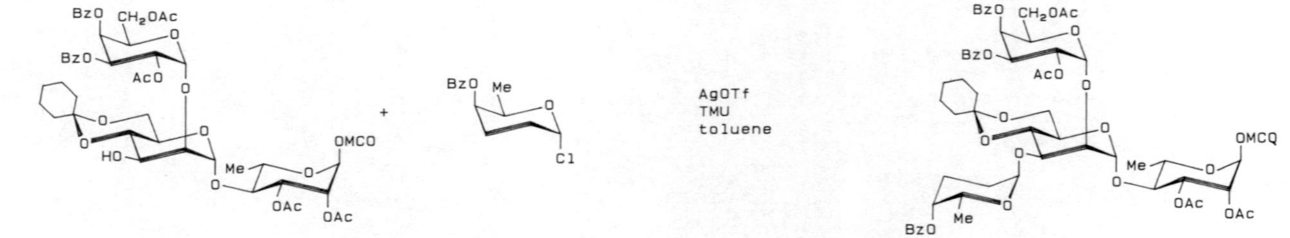

30%; $[\alpha]_D$ +64 (chloroform); C-13; Ref.: 109

α-D-Galp-(1→2)
3,4,6-trideoxy-α-D-erythro-Hexp-(1→3) 〉α-D-Manp-(1→4)-L-Rha

α-p-(1→OMCO); $[\alpha]_D$ +84 (water); C-13; Ref.: 109

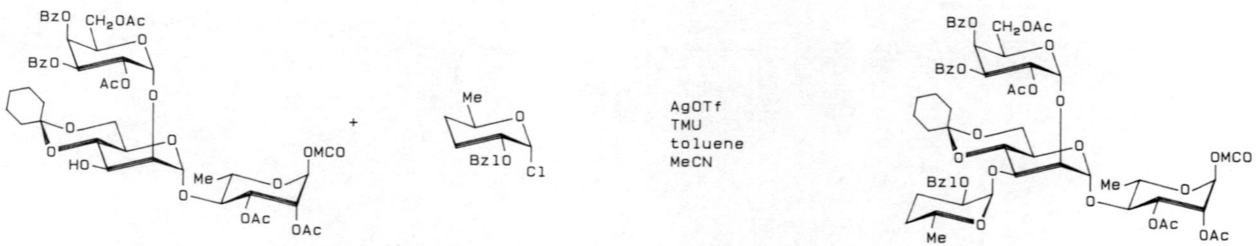

58%; $[\alpha]_D$ +95 (chloroform); Ref.: 109

α-D-Galp-(1→2)
3,4,6-trideoxy-β-D-erythro-Hexp-(1→3) 〉α-D-Manp-(1→4)-L-Rha

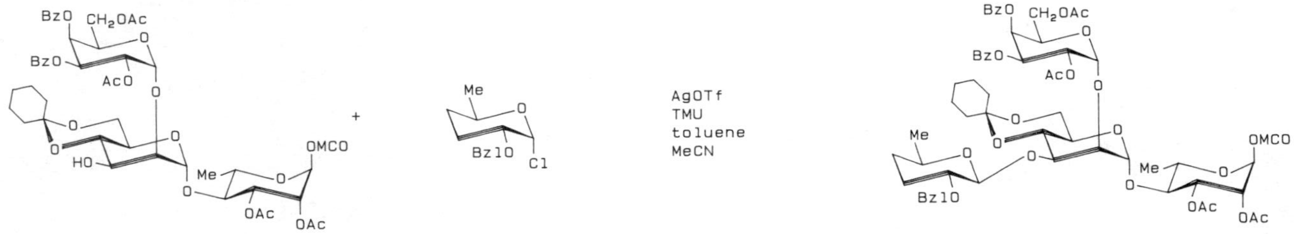

28%; H-1; Ref.: 109

α-L-Rhap-(1→2)-α-L-Rhap-(1→3)-α-L-Rhap-(1→2)-L-Rha

α-p-(1-OMCO); $[\alpha]_D$ -68.2 (water); C-13; Ref.: 111

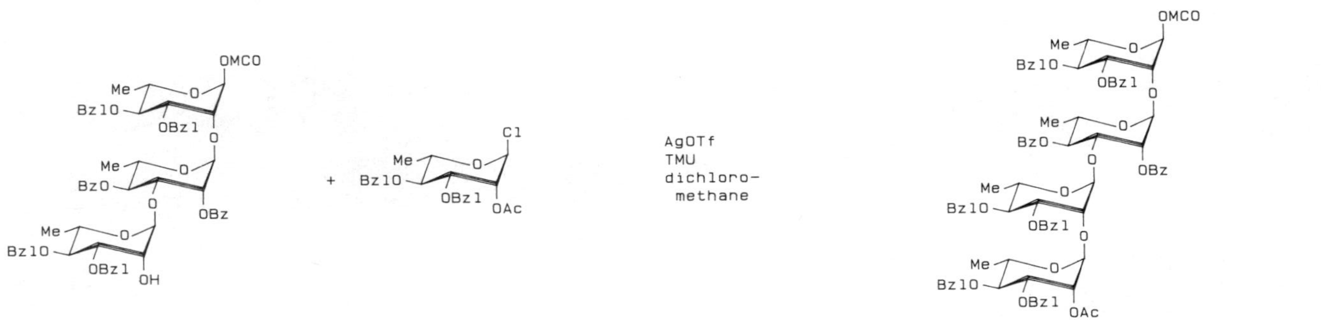

85%; $[\alpha]_D$ +18.9 (chloroform); C-13; Ref.: 111

β-D-Galp-(1→3)-α-L-Rhap-(1→3)-α-L-Rhap-(1→2)-L-Rha

$[\alpha]_D$ -30 (water); C-13; Ref.: 112

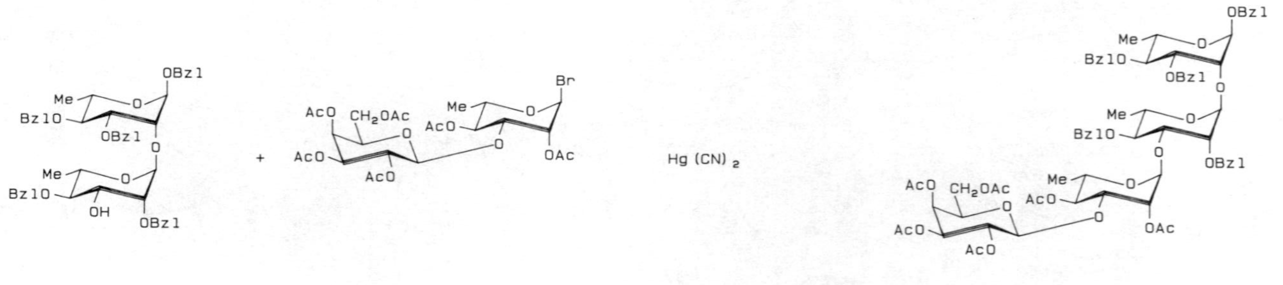

$[\alpha]_D$ -20 (chloroform); C-13; Ref.: 112

3-O-Me-β-L-Xylp-(1→4)-α-L-Rhap-(1→4)-α-L-Rhap-(1→2)-L-Rha

amorphous; $[\alpha]_D$ -22 → -18 (water); C-13; Ref.: 113

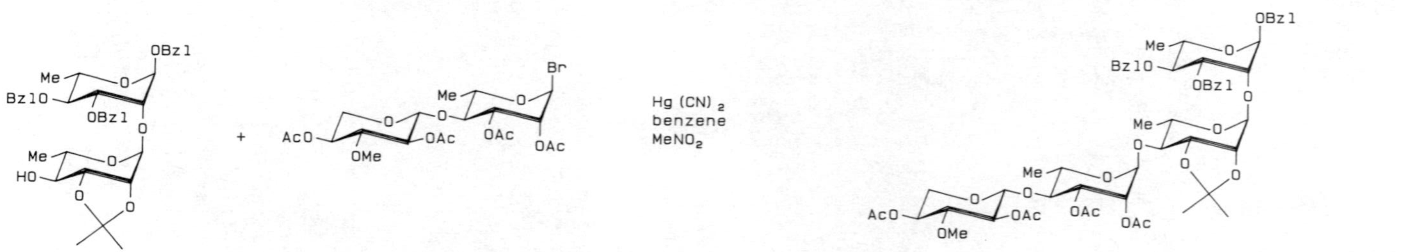

38.9%; foam; $[\alpha]_D$ -29 (chloroform); C-13; Ref.: 113

β-D-GlcpNAc-(1→2)
α-D-Glcp-(1→3)
⟩α-L-Rhap-(1→2)-L-Rha

α-p-(1→OMe); amorphous powder; $[\alpha]_D$ +46 (MeOH); C-13; Ref.: 114

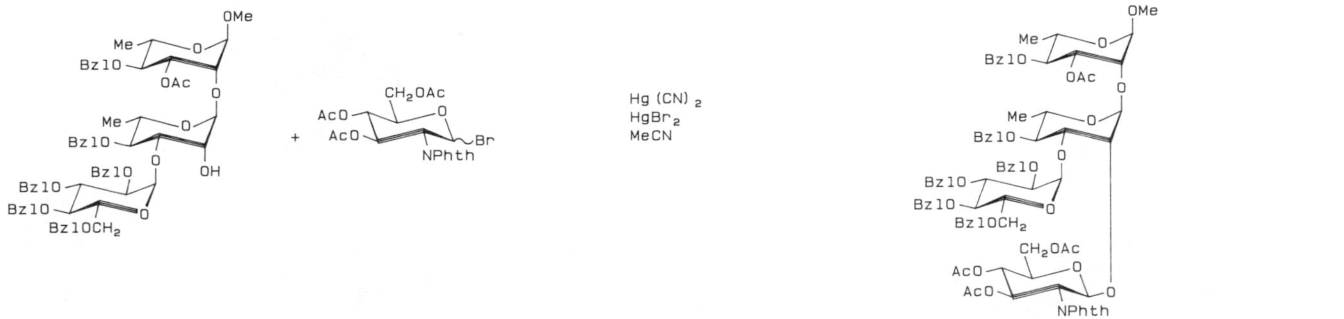

69%; syrup; $[\alpha]_D$ +71 (chloroform); C-13; Ref.: 114

β-D-GlcpNAc-(1→2)-α-L-Rhap-(1→2)-α-L-Rhap-(1→3)-L-Rha

α-p-(1→OMCO); $[\alpha]_{589}$ −54 (MeOH); C-13; Ref.: 115

α-p-[1→O$(CH_2)_8$CONH-NH_2]; $[\alpha]_{589}$ −51 (MeOH); C-13; Ref.: 115

α-p-(1→OMe); $[\alpha]_D$ −64 (water); C-13; Ref.: 116

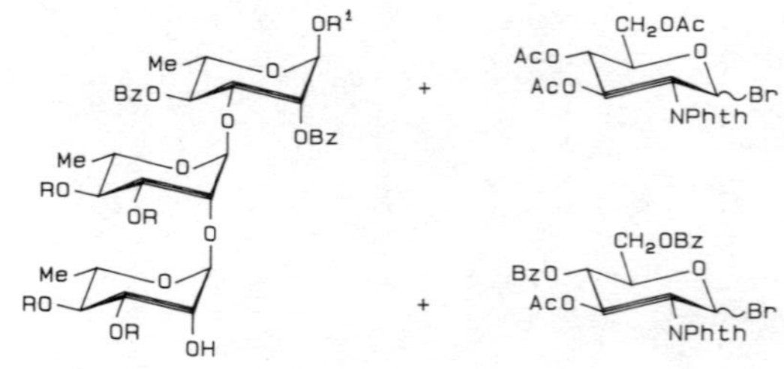

I AgOTf
s-Coll
dichloro-
methane
R^1=MCO; R=Bzl

II $Hg(CN)_2$
$HgBr_2$
MeCN
R^1=Me; R=Bz

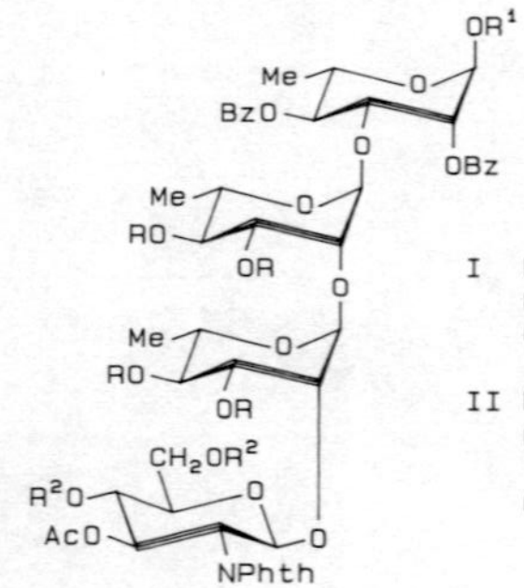

I R^1=MCO; R=Bzl; R^2=Ac
85%; $[\alpha]_D$ +48.4 (chloroform)
C-13; Ref.: 115

II R^1=Me; R=R^2=Bz
68%; m.p. 152–154 °C;
$[\alpha]_D$ +123 (chloroform)
C-13; Ref.: 116

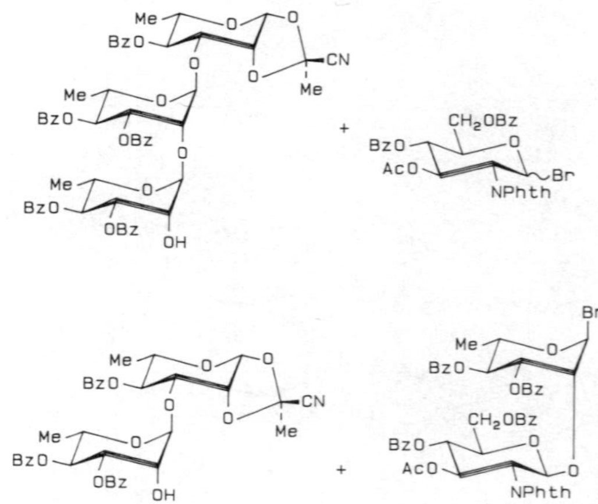

I

II

II

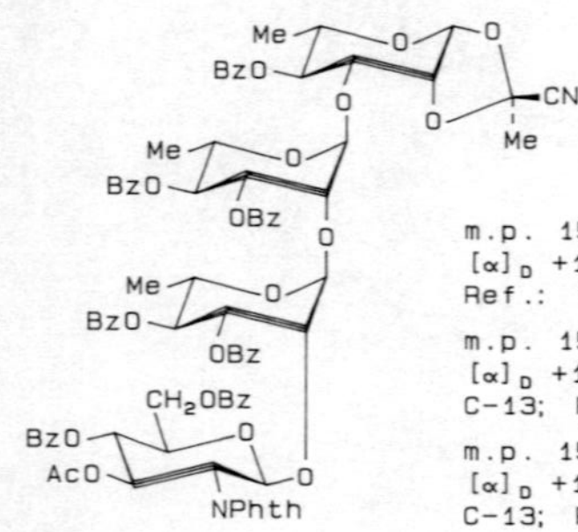

m.p. 154–158 °C;
$[\alpha]_D$ +106 (chloroform)
Ref.: 118, 119

m.p. 154–157 °C;
$[\alpha]_D$ +105.5 (chloroform)
C-13; Ref.: 118, 119

m.p. 154–158
$[\alpha]_D$ +105.5 (chloroform)
C-13; Ref.: 117

I 14%; Ref.: 118, 119

II 45.5%; Ref.: 117, 118, 119

48%; Ref.: 118, 119

α-L-Rhap-(1→2)
α-D-Glcp-(1→3) 〉 α-L-Rhap-(1→3)-L-Rha

amorphous powder; $[\alpha]_D$ -11 (MeOH); -5 (water); C-13; Ref.: 120

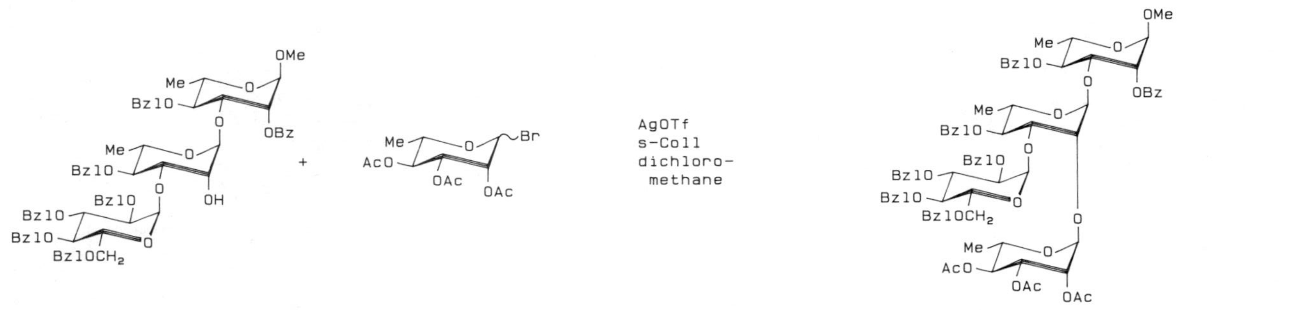

$[\alpha]_D$ +5.3 (chloroform); C-13; Ref.: 120

β-D-GalpNAc-(1→2)
β-L-Rhap-(1→4) 〉 β-L-Rhap-(1→2)-L-Rha

$[\alpha]_D$ +4 (MeOH); Ref.: 121

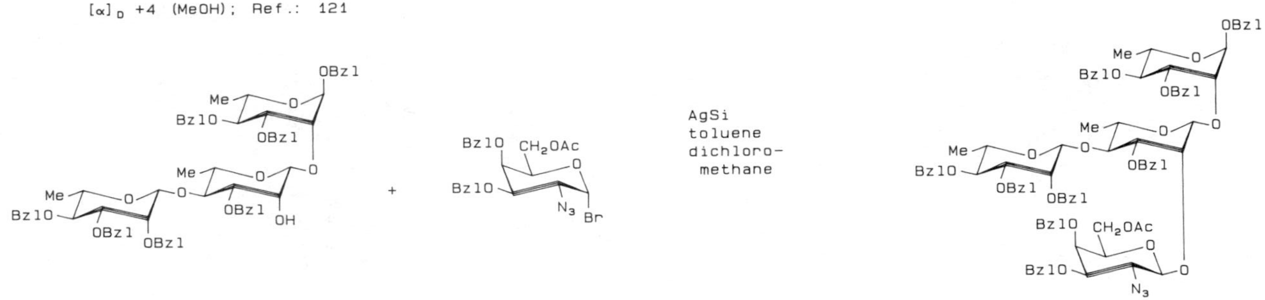

50%; $[\alpha]_D$ +1 (chloroform); Ref.: 121

β-D-GalpNAc-(1→2)
β-L-Rhap-(1→4) 〉β-L-Rhap-(1→4)-L-Rha

$[\alpha]_D$ +12.0 (water); Ref.: 122

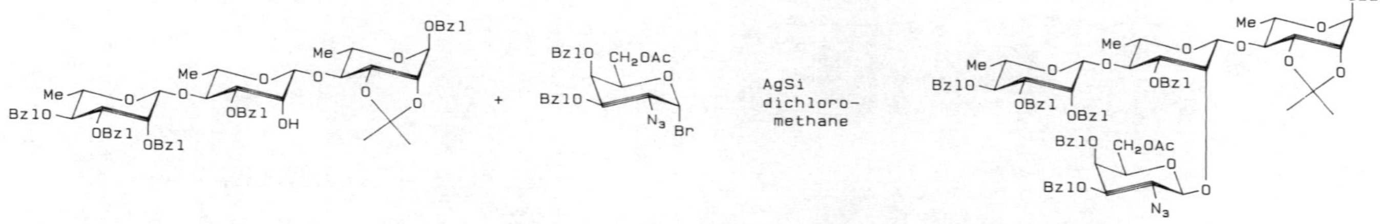

60%; syrup; $[\alpha]_D$ −10.4 (chloroform); Ref.: 122

α-D-GlcpN-(1→3)
β-D-Manp-(1→4) 〉α-D-Galp-(1→4)-L-Rha

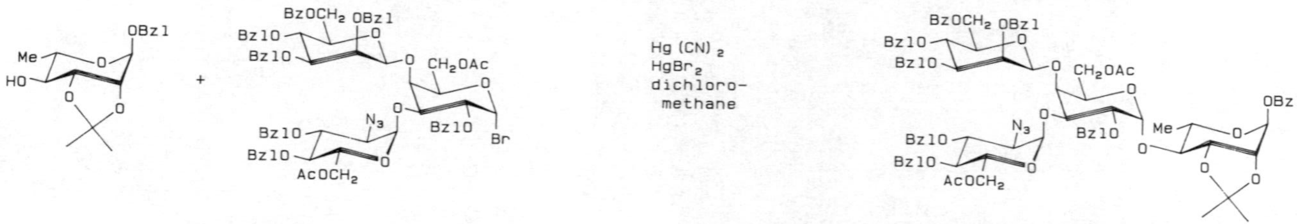

32%; $[\alpha]_D$ +36.1 (dichloromethane); Ref.: 123

β-D-GlcpNAc-(1→3)
β-D-Manp-(1→4) ⟩ α-D-Galp-(1→4)-L-Rha

$[\alpha]_D$ +39.6 (water); Ref.: 124

α-p-(1→OMCO); $[\alpha]_D$ +14.9 (MeOH); Ref.: 124

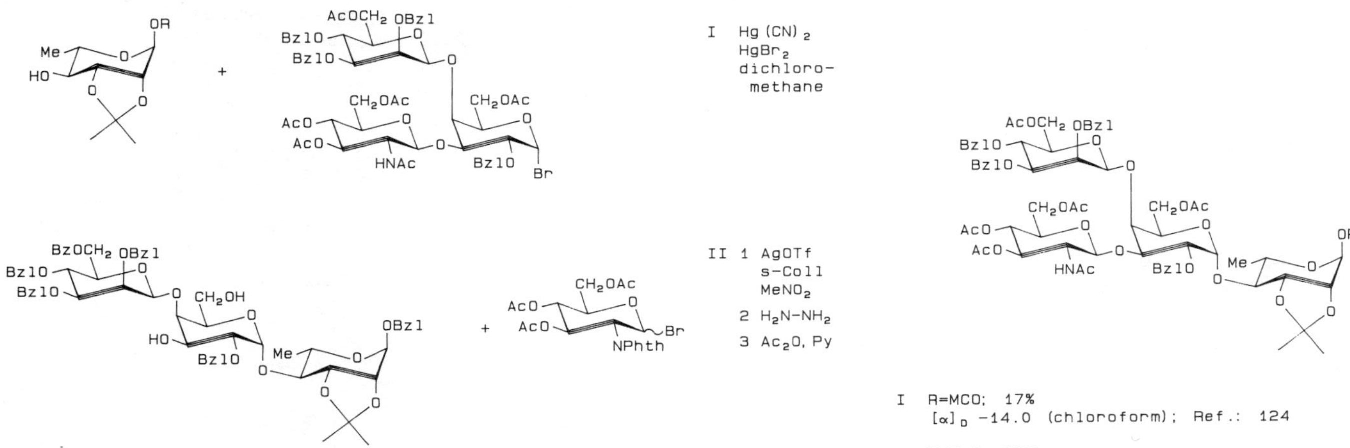

I R=MCO; 17%
$[\alpha]_D$ −14.0 (chloroform); Ref.: 124

R=Bzl; 20%
$[\alpha]_D$ −16.8 (chloroform); Ref.: 124

II R=Bzl; 59%
$[\alpha]_D$ −16.6 (chloroform); Ref.: 124

β-D-GlcpNAc-(1→3)
β-D-Manp-(1→4) 〉β-D-Galp-(1→4)-L-Rha

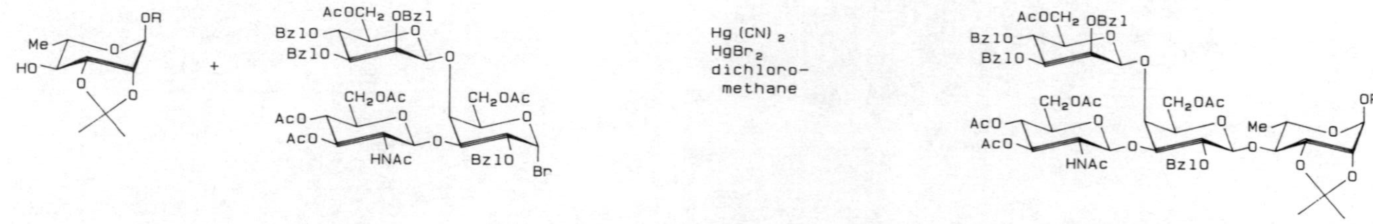

R=MCO; 13%
$[\alpha]_D$ −30.2 (chloroform); Ref.: 124

R=Bzl; 17%
$[\alpha]_D$ −34.9 (chloroform); Ref.: 124

α-D-GlcpA-(1→6)-α-D-Glcp-(1→2)
α-L-Rhap-(1→3) ⟩ L-Rha

α-p-(1→OMe); m.p. 156–160 °C; $[\alpha]_D$ +56 (water); Ref.: 125

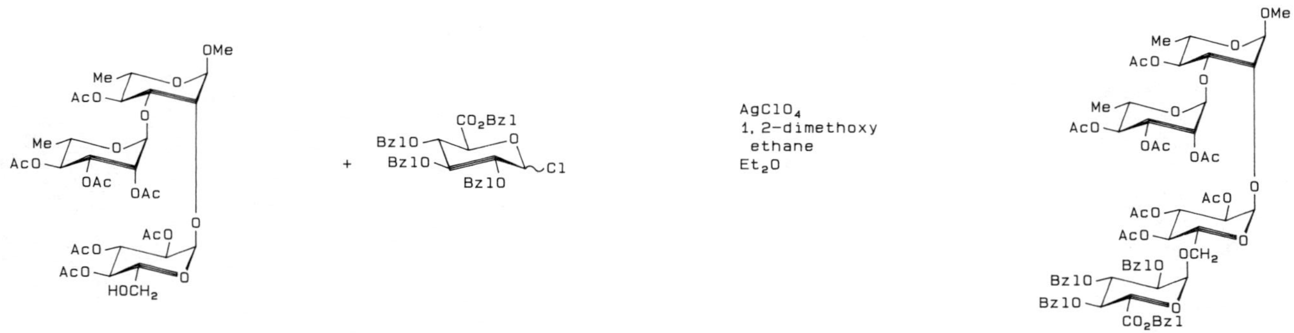

m.p. 173–174 °C; $[\alpha]_D$ +57 (chloroform); Ref.: 125

β-D-GlcpA-(1→6)-α-D-Glcp-(1→2) \
α-L-Rhap-(1→3) / L-Rha

α-p-(1→OMe); m.p. 169–172 °C; $[\alpha]_D$ +5 (water); Ref.: 125

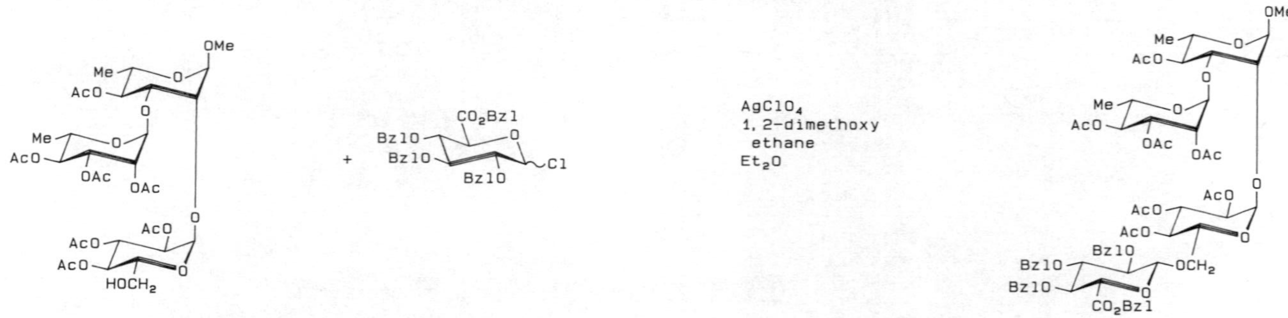

m.p. 77.6–78.8 °C; $[\alpha]_D$ +32 (chloroform); Ref.: 125

α-D-Glcp-(1→3)-α-L-Rhap-(1→2) \
                                                       ⟩ L-Rha
α-D-Glcp-(1→3) /

α-p-(1→OMe); amorphous; $[\alpha]_D$ +81 (MeOH); C-13; Ref.: 126, 127

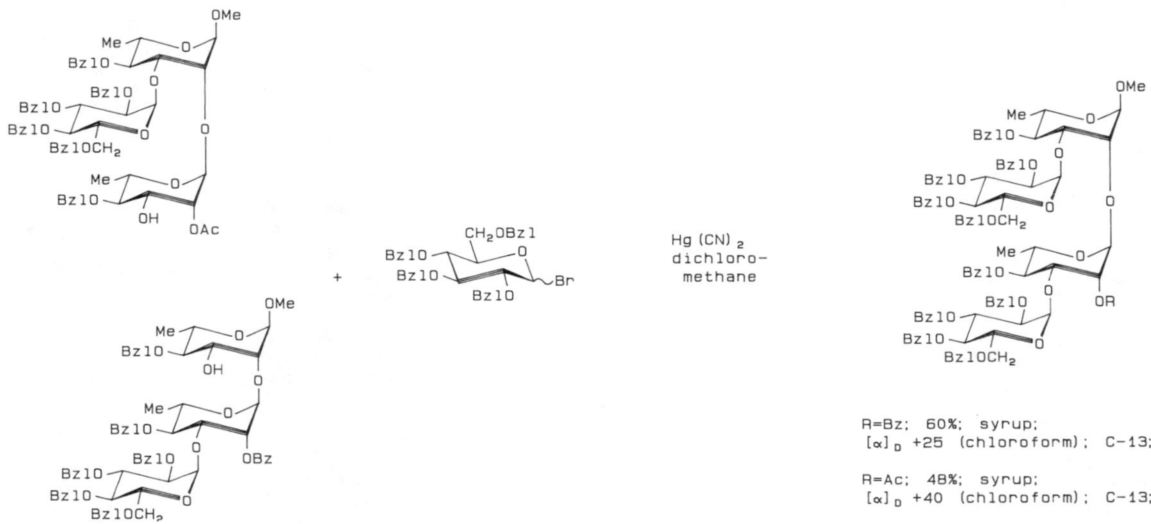

R=Bz; 60%; syrup;
$[\alpha]_D$ +25 (chloroform); C-13; Ref.: 126, 127

R=Ac; 48%; syrup;
$[\alpha]_D$ +40 (chloroform); C-13; Ref.: 126, 127

β-D-GlcpNAc-(1→2)-α-L-Rhap-(1→2)
α-D-Glcp-(1→3) 〉L-Rha

α-p-(1→OMe); $[\alpha]_D$ +20.2 (MeOH); C-13; Ref.: 128

α-p-(1→OMe); amorphous powder; $[\alpha]_D$ +31 (MeOH); +33 (water); C-13; Ref.: 120

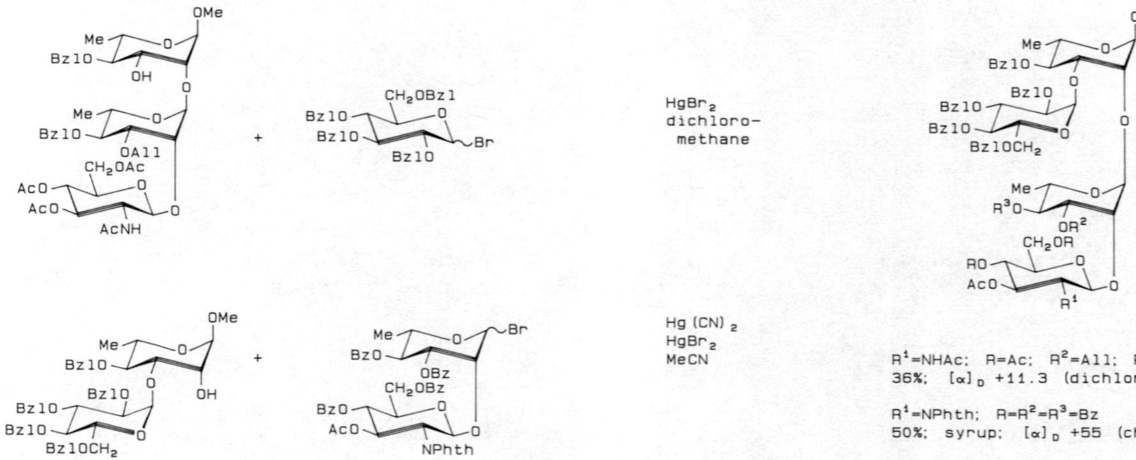

R^1=NHAc; R=Ac; R^2=All; R^3=Bzl
36%; $[\alpha]_D$ +11.3 (dichloromethane); C-13; Ref.: 128

R^1=NPhth; R=R^2=R^3=Bz
50%; syrup; $[\alpha]_D$ +55 (chloroform); C-13; Ref.: 120

β-D-GlcpNAc-(1→2)-α-L-Rhap-(1→2) ⟩ L-Rha
β-D-Glcp-(1→3)

7%; $[\alpha]_D$ -1.0 (dichloromethane); C-13; Ref.: 128

α-D-Glcp-(1→4)-α-D-Glcp-(1→3) ⟩ L-Rha
β-D-ManpNAc-(1→4)

syrup; $[\alpha]_D$ +68.5 (water); Ref.: 107, 129

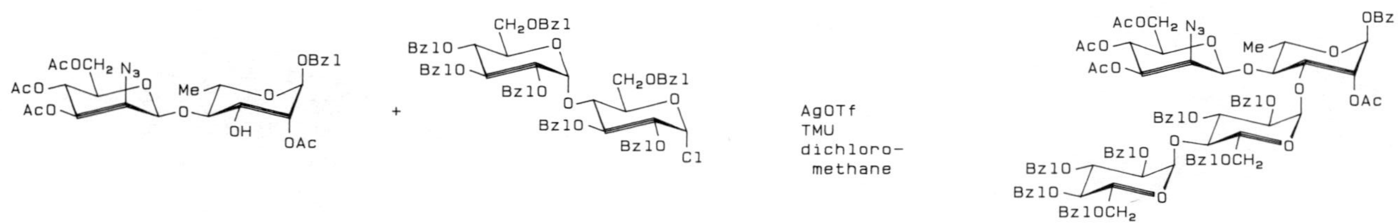

54%; $[\alpha]_D$ -2.3 (chloroform); Ref.: 107, 129

α-D-Glcp-(1→4)-β-D-Glcp-(1→3) 〉L-Rha
β-D-ManpN-(1→4)

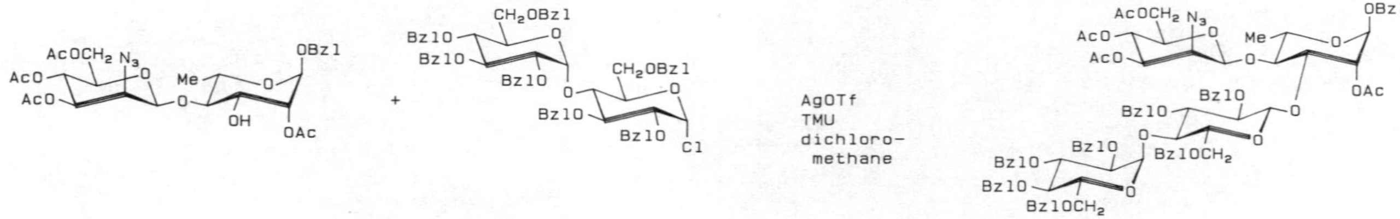

25%; $[\alpha]_D$ −17.2 (chloroform); Ref.: 129

β-D-Glcp-(1→2) 〉β-D-Glcp-(1→4)-D-Gal
β-D-Xylp-(1→3)

m.p. 185–189 °C (dec.) after foaming at 177–180 °C; $[\alpha]_D$ +2.0 (water); Ref.: 130

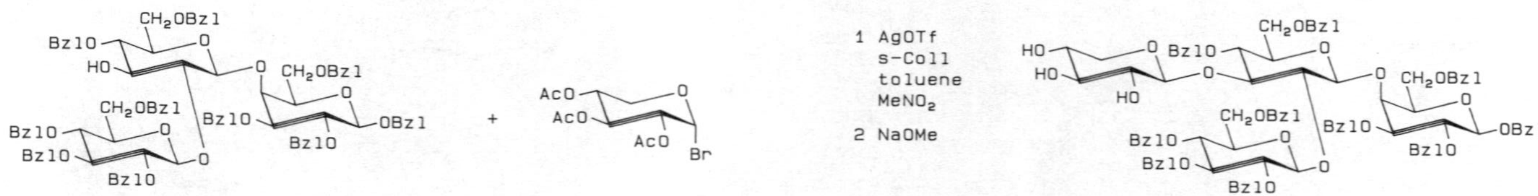

76%; $[\alpha]_D$ +3.9 (chloroform); Ref.: 130

α-Parp-(1→3)-α-D-Manp-(1→4)-α-L-Rhap-(1→3)-D-Gal

α-p-(1→O-Trifluoroacetamidophenyl); amorphous solid; $[\alpha]_D$ +119 (water); C-13; Ref.: 131

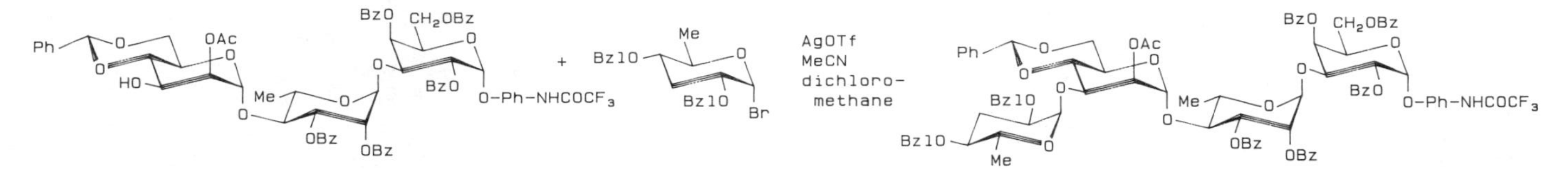

Parp = 3,6-Dideoxy-D-ribo-hexopyranosyl

54%; m.p. 193-196 °C; $[\alpha]_D$ +121 (chloroform); C-13; Ref.: 131

β-Parp-(1→3)-α-D-Manp-(1→4)-α-L-Rhap-(1→3)-D-Gal

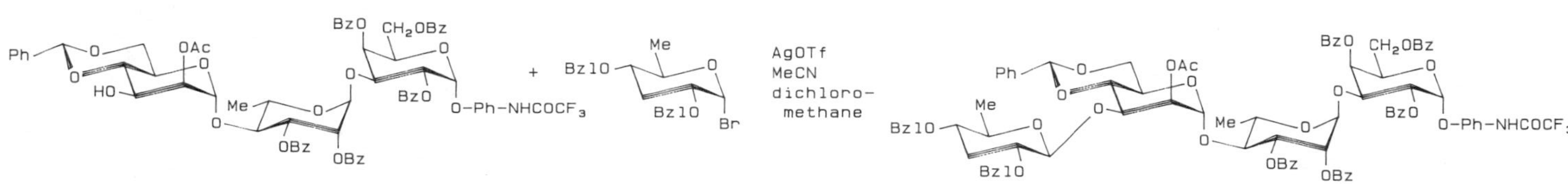

Parp = 3,6-Dideoxy-D-ribo-hexopyranosyl

23%; $[\alpha]_D$ +95 (chloroform); C-13; Ref.: 131

α-Tyvp-(1→3)-α-D-Manp-(1→4)-α-L-Rhap-(1→3)-D-Gal

α-p-(1→O-Trifluoroacetamidophenyl); amorphous solid; $[\alpha]_D$ +107 (water); C-13; Ref.: 131

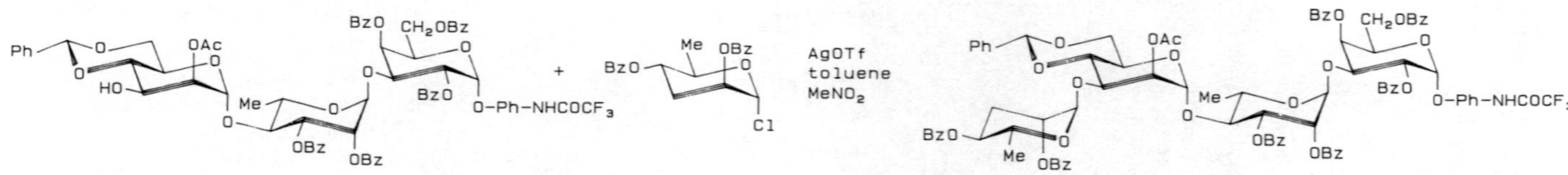

Tyvp = 3,6-Dideoxy-D-arabino-hexopyranosyl

66%; m.p. 238-240 °C; $[\alpha]_D$ +71 (chloroform); C-13; Ref.: 131

α-Abep-(1→3)-α-D-Manp-(1→4)-α-L-Rhap-(1→3)-D-Gal

α-p-(1→O-Trifluoroacetamidophenyl); amorphous solid; $[\alpha]_D$ +100 (water); C-13; Ref.: 131

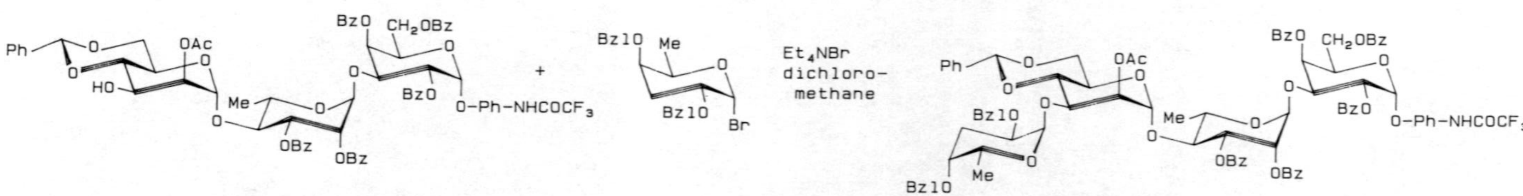

61%; m.p. 187-188 °C; $[\alpha]_D$ +103 (chloroform); C-13; Ref.: 131

β-D-Galp-(1→3)-β-D-Galp-(1→3)-β-D-Galp-(1→3)-D-Gal

β-p-(1→OMe); m.p. 215-218 °C; [α]$_D$ +33.3 (water); C-13; Ref.: 132

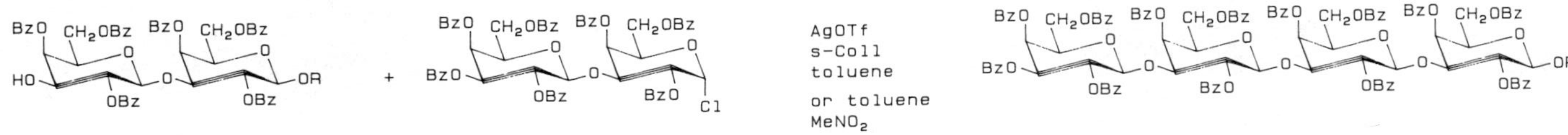

R=Ac; 54%; [α]$_D$ +65 (chloroform); C-13; Ref.: 132

R=Me; 61.4%; [α]$_D$ +57 (chloroform); C-13; Ref.: 132

β-D-Glcp-(1→2)-β-D-Galp-(1→2)-β-D-Galp-(1→6)-D-Gal

C-13; Ref.: 133

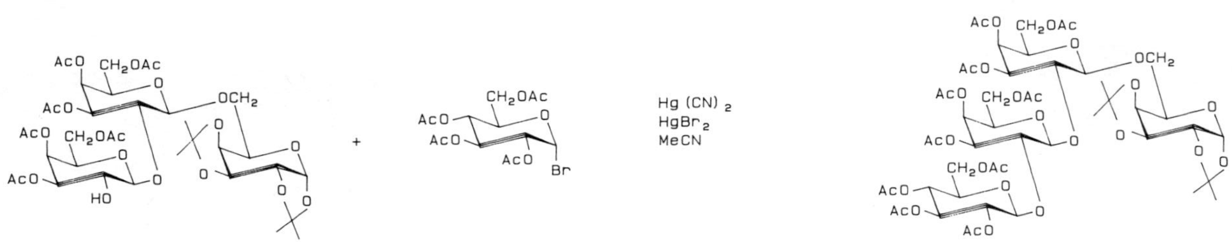

52%; [α]$_D$ +54 (chloroform); C-13; Ref.: 133

α-D-Galp-(1→6)-β-D-Galp-(1→6)-β-D-Galp-(1→6)-D-Gal

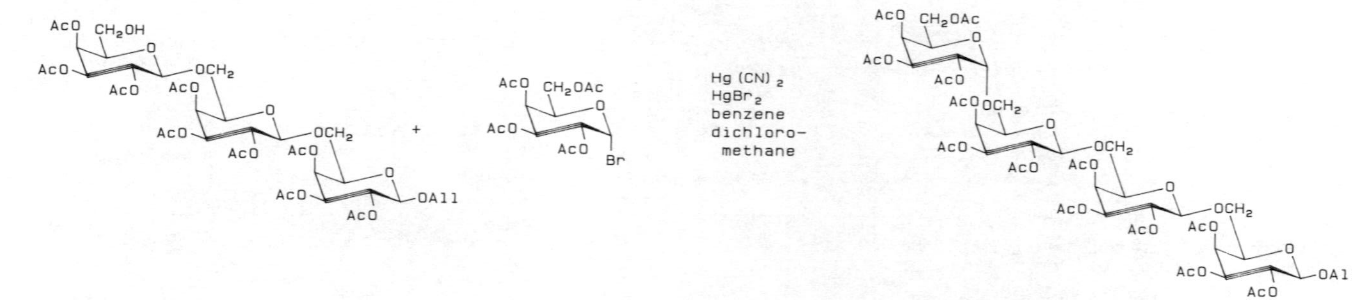

4%; $[\alpha]_D$ +26.4 (chloroform); C-13; Ref.: 134

β-D-Galp-(1→6)-β-D-Galp-(1→6)-β-D-Galp-(1→6)-D-Gal

β-p-(1→OPr); $[\alpha]_D$ +37.2 (water); C-13; Ref.: 135

β-p-(1→O-CH2-epoxide); $[\alpha]_D$ −11.6 (water); C-13; Ref.: 134

β-p-(1→OMe); m.p. 180–185 °C; $[\alpha]_D$ −10.3 (water); C-13; Ref.: 136

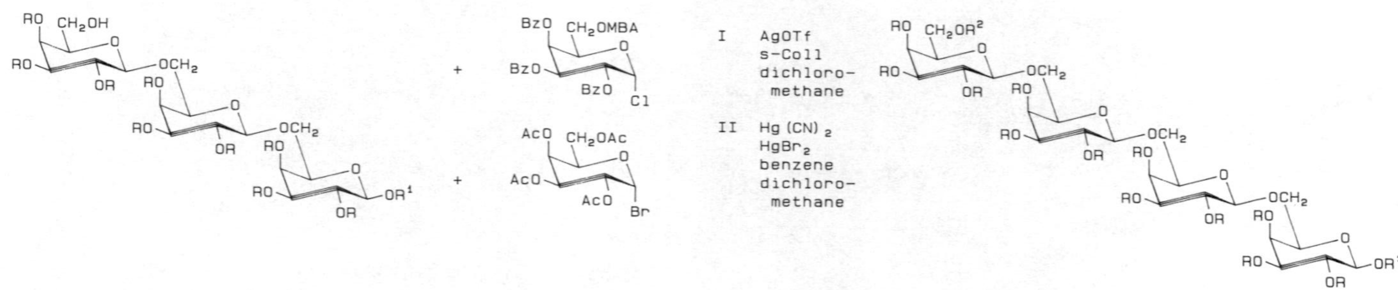

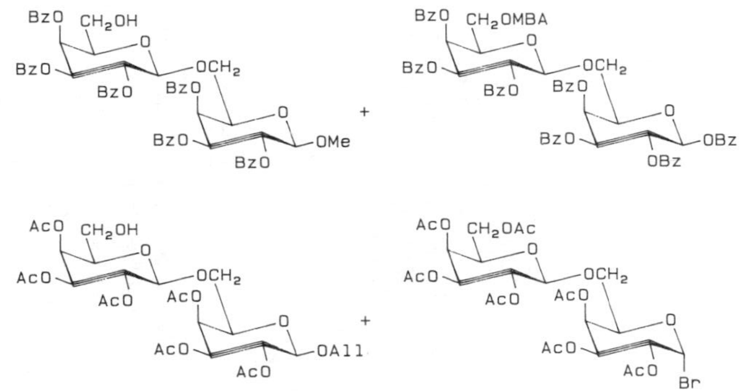

IV TMSOTf dichloromethane

I R^1=Me; R=Bz; R^2=MBA; 78%; m.p. 273-275 °C; $[\alpha]_D$ +104 (chloroform); C-13; Ref.: 136, 137

II R^1All; R=R^2=Ac; 70% $[\alpha]_D$ -22 (chloroform); C-13; Ref.: 134

IV R^1=Me; R=Bz; R^2=MBA 72%; Ref.: 136, 137

V R^1=All; R=R^2=Ac; 68%; Ref.: 134

V Hg(CN)$_2$ $HgBr_2$ dichloromethane

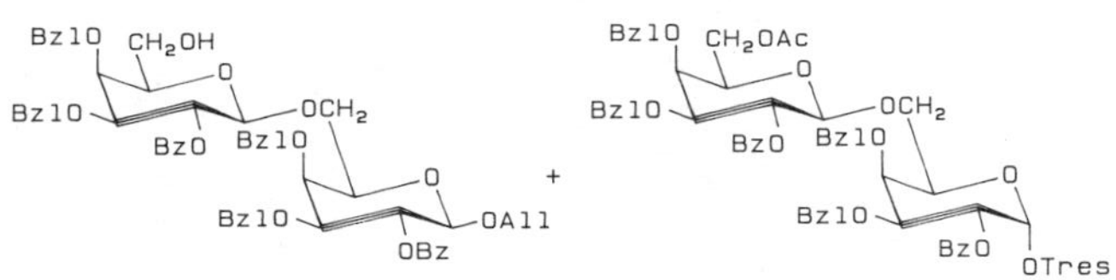

III MeCN

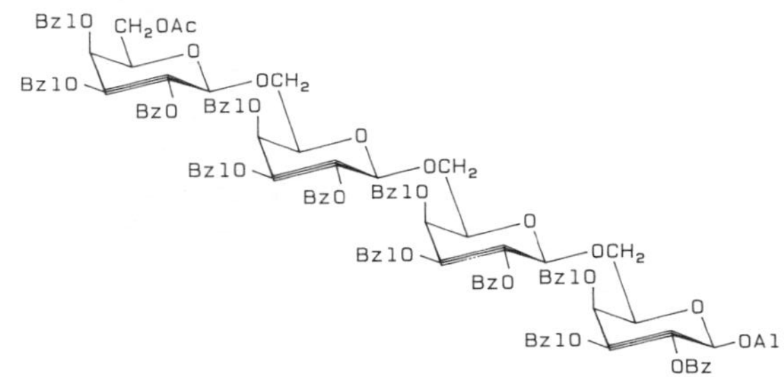

III 77%; $[\alpha]_D$ +25.8 (chloroform); C-13; Ref.: 135

β-D-Manp-(1→4)-α-L-Rhap-(1→3)
α-D-Glcp-(1→4)
⟩D-Gal

$[\alpha]_D$ +35.1 (water); C-13; Ref.: 138

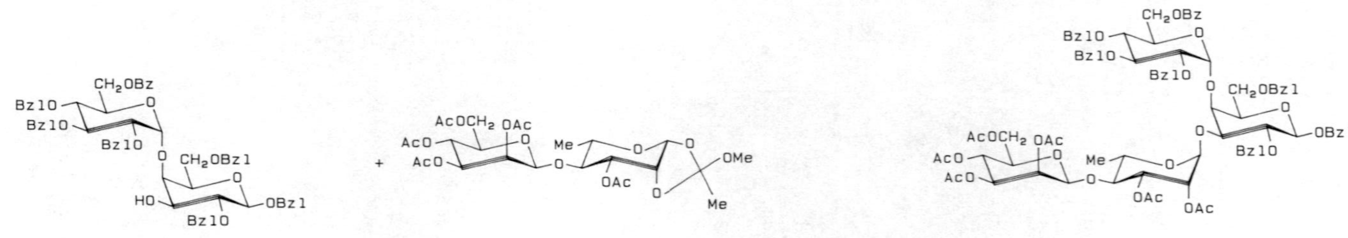

35%; $[\alpha]_D$ +15.5 (chloroform); Ref.: 138

β-D-Glcp-(1→4)-α-L-Rhap-(1→3)
α-D-Glcp-(1→6)
⟩D-Gal

$[\alpha]_D$ +32 (water); Ref.: 139

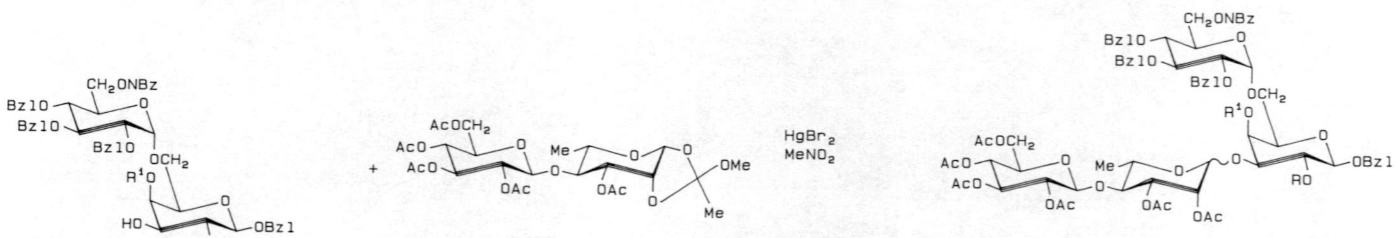

R=Ac; R¹=H; 52%; $[\alpha]_D$ −12 (chloroform); Ref.: 139

R=R¹=Bzl; ~α; 40%
$[\alpha]_D$ −2.7 (chloroform); Ref.: 139

α-D-Manp-(1→4)-α-L-Rhap-(1→3) \ D-Gal
α-D-Glcp-(1→6) /

[α]$_D$ +39 (water); Ref.: 140

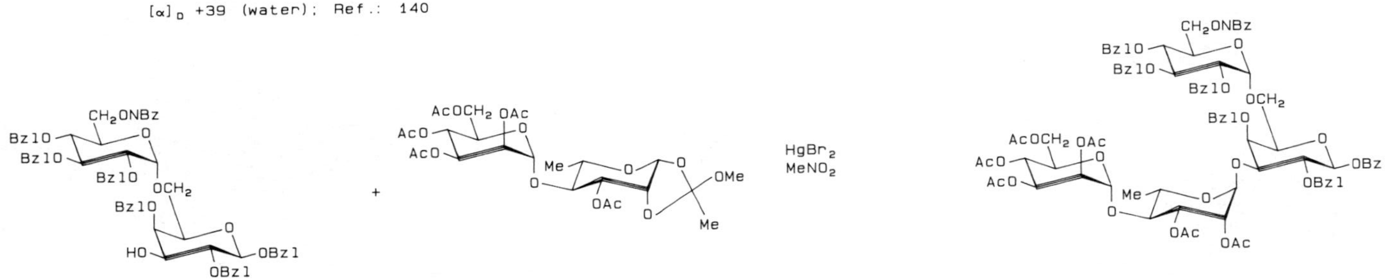

27%; [α]$_D$ +24 (chloroform); Ref.: 140

β-D-Manp-(1→4)-α-L-Rhap-(1→3) \ D-Gal
α-D-Glcp-(1→6) /

[α]$_D$ +27 (water); Ref.: 140

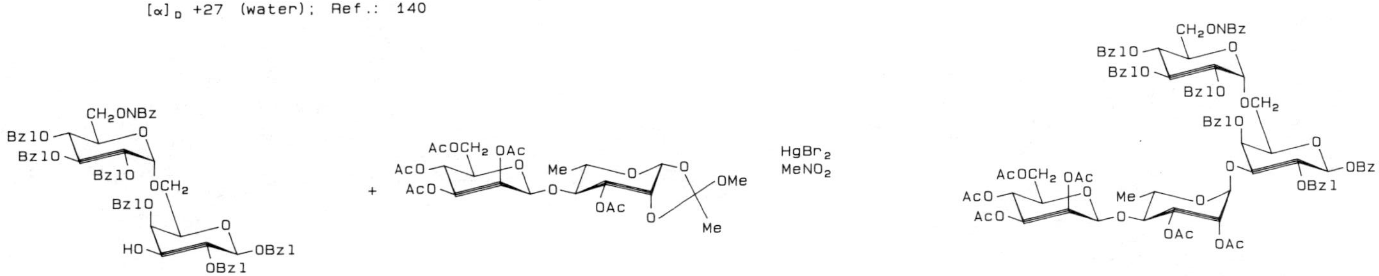

27%; [α]$_D$ -0.075 (chloroform); Ref.: 140

β-D-Glcp-(1→4)-β-L-Rhap-(1→3)-[α-D-Glcp-(1→6)]-D-Gal

$[\alpha]_D$ +60 (water); Ref.: 139

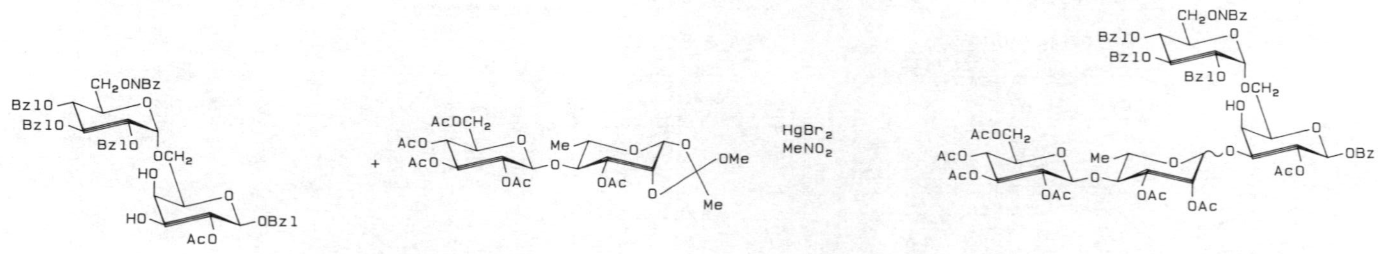

52%; ~β; $[\alpha]_D$ −12 (chloroform); Ref.: 139

β-D-Manp-(1→4)-β-L-Rhap-(1→3)-[α-D-Glcp-(1→6)]-D-Gal

$[\alpha]_D$ +56 (water); Ref.: 140

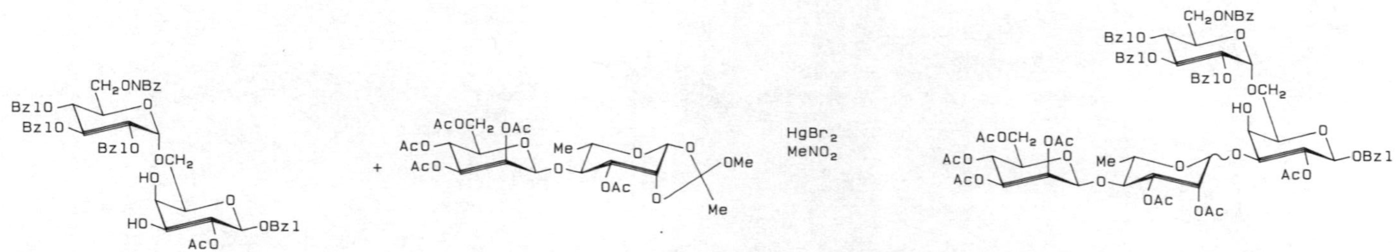

40%; ~β; $[\alpha]_D$ −9.5 (chloroform); Ref.: 140

β-D-Glcp-(1→2)-β-D-Glcp-(1→2) \
β-D-Galp-(1→6) / D-Gal

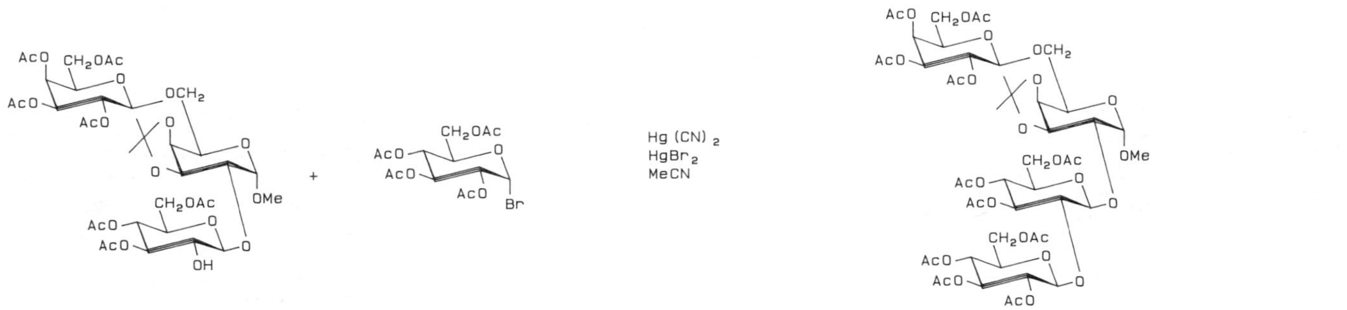

65%; glass; $[\alpha]_D$ +33.4 (chloroform); C-13; Ref.: 141

β-D-Glcp-(1→6)-β-D-Galp-(1→6) \
β-D-Glcp-(1→2) / D-Gal

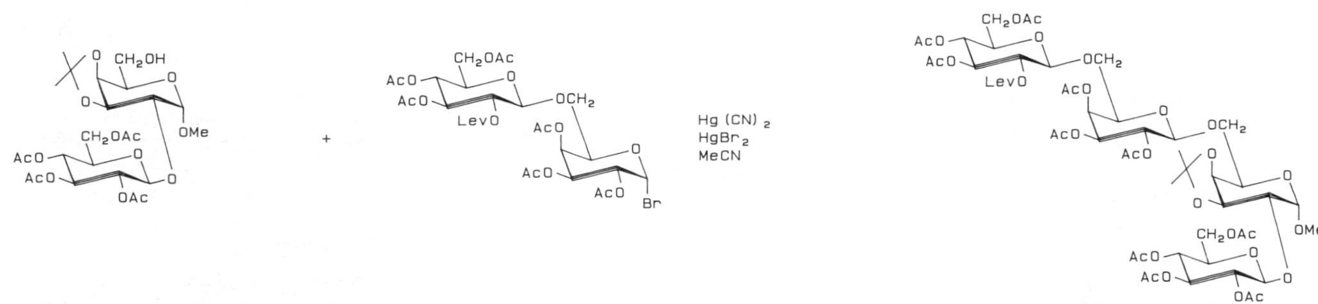

31%; glass; Ref.: 133

β-D-Galp-(1→3)-β-D-GlcpNAc-(1→3)
β-D-GlcpNAc-(1→6) 〉D-Gal

trihydrate; 50% overall yield; m.p. 186–188 °C; $[\alpha]_D$ +2.5 (water); Ref.: 142

m.p. 186–188 °C; $[\alpha]_D$ +2.5 (water); Ref.: 143

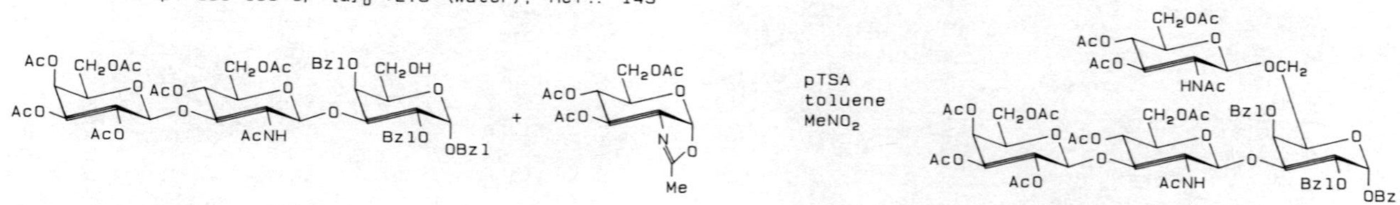

78%; m.p. 116–118 °C; $[\alpha]_D$ +5.2 (chloroform); Ref.: 143

β-D-Galp-(1→4)-β-D-GlcpNAc-(1→6)
β-D-GlcpNAc-(1→3) 〉D-GalNAc

$[\alpha]_{578}$ −5 (water); Ref.: 144, 145

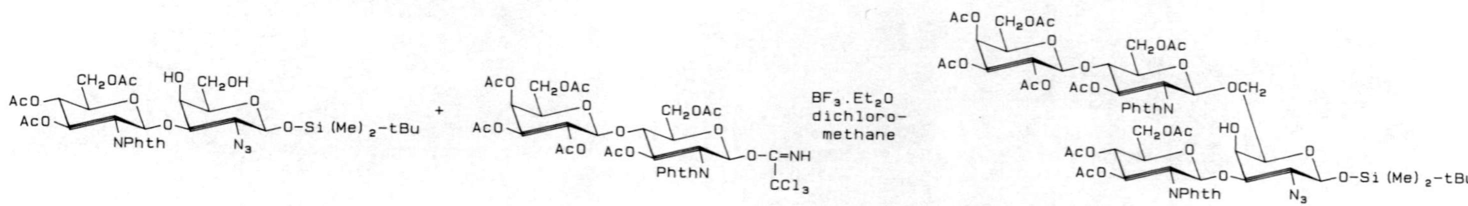

71%; m.p. 168–170 °C; $[\alpha]_{578}$ +10 (chloroform); Ref.: 144, 145

β-D-GlcpNAc-(1→6)-β-D-Galp-(1→3)
β-D-GlcpNAc-(1→6) 〉 D-GalNAc

α-p-(1→OBzl); m.p. 213-216 °C; $[\alpha]_D$ +40.4 (water); C-13; Ref.: 146

51.6%; $[\alpha]_D$ +31.2 (MeOH); Ref.: 146

β-D-Galp-(1→6)-β-D-Galp-(1→6)-β-D-Galp-(1→6)-3-deoxy-3-fluoro-D-Gal

β-p-(1→OMe); m.p. 254-256 °C; $[\alpha]_D$ −10.6 (water); Ref.: 147

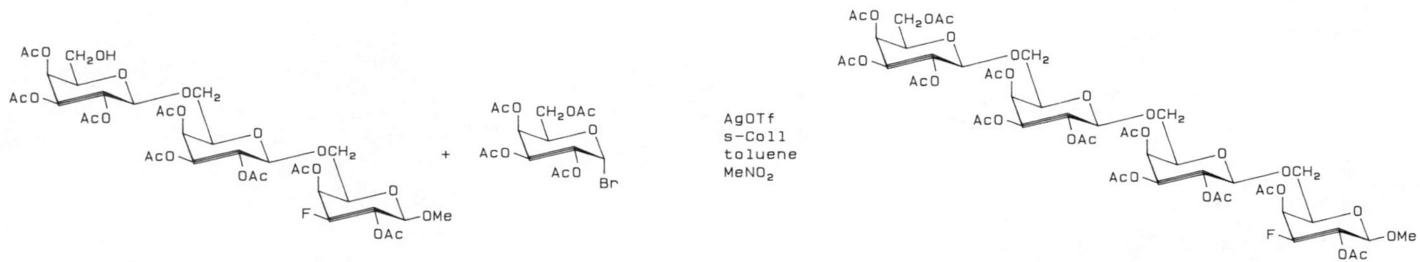

44%; glassy solid; C-13; Ref.: 147

2-O-Me-α-L-Fucp-(1→4)-2-O-Me-α-L-Fucp-(1→3)-α-L-Rhap-(1→2)-6-deoxy-L-Tal

amorphous solid; $[\alpha]_D$ -176 (water); Ref.: 148

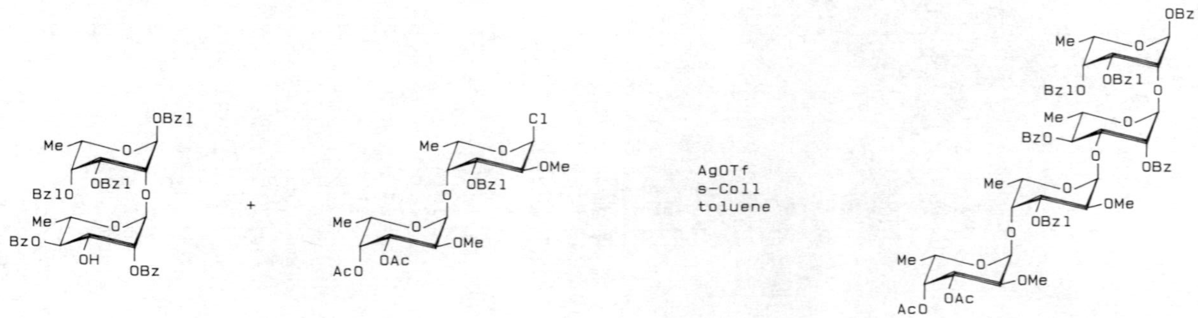

82%; $[\alpha]_D$ -79 (chloroform); Ref.: 148

2,3-di-O-Me-α-L-Fucp-(1→4)-2,3-di-O-Me-α-L-Fucp-(1→3)-α-L-Rhap-(1→2)-6-deoxy-L-Tal

amorphous solid; $[\alpha]_D$ -177 (water); Ref.: 148

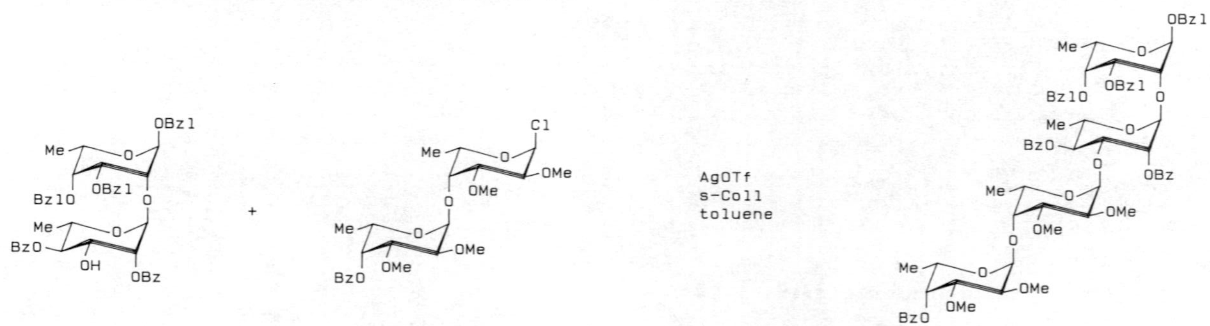

78%; $[\alpha]_D$ -75 (chloroform); Ref.: 148

β-D-Glcp-(1→4)-2,6-dideoxy-2-I-α-D-Altp-(1→3)-2,6-dideoxy-α-D-ribo-Hexp-(1→4)-2,6-dideoxy-D-ribo-Hexp

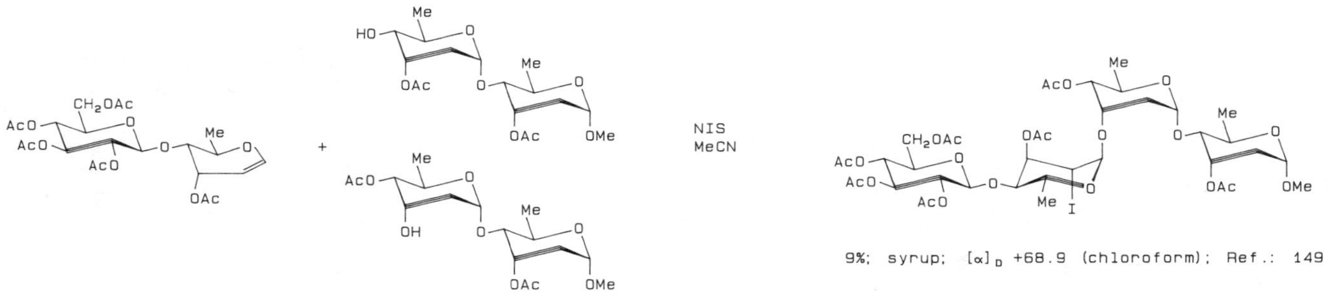

9%; syrup; $[\alpha]_D$ +68.9 (chloroform); Ref.: 149

β-D-Glcp-(1→4)-2,6-dideoxy-α-D-ribo-Hexp-(1→4)-2,6-dideoxy-α-D-ribo-Hexp-(1→4)-2,6-dideoxy-D-ribo-Hexp

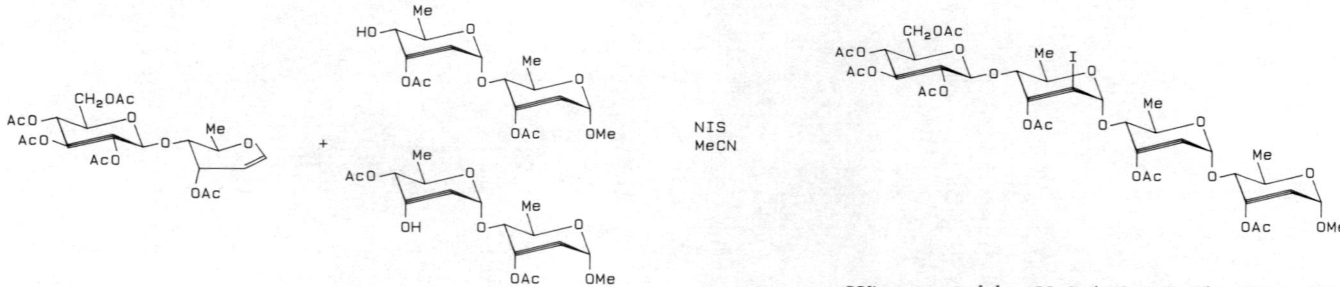

22%; syrup; $[\alpha]_D$ +88.2 (chloroform); Ref.: 149

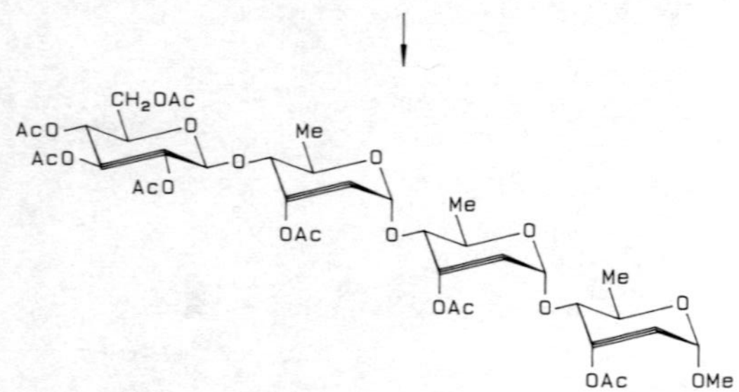

36%; syrup; $[\alpha]_D$ +92 (chloroform); Ref.: 149

α-D-Galp-(1→6)-α-D-Galp-(1→6)-α-D-Glcp-(1→2)-D-Fru

m.p. 122–123 °C; $[\alpha]_D$ +140.2 (water); Ref.: 150

monohydrate; $[\alpha]_D$ +147.3 (water); Ref.: 150

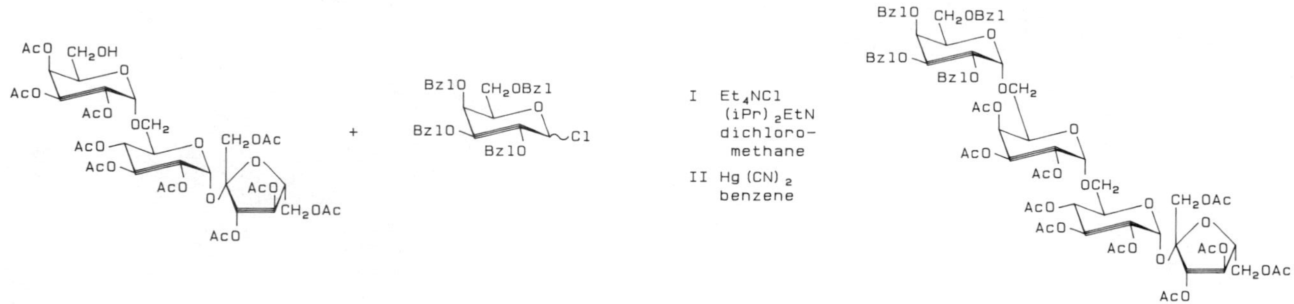

I 32%; m.p. 70–72 °C; $[\alpha]_D$ +91.9 (chloroform); Ref.: 150

II 45%; m.p. 66–68 °C; $[\alpha]_D$ +88.2 (chloroform); Ref.: 150

β-D-Galp-(1→6)-α-D-Galp-(1→6)-α-D-Glcp-(1→2)-D-Fru

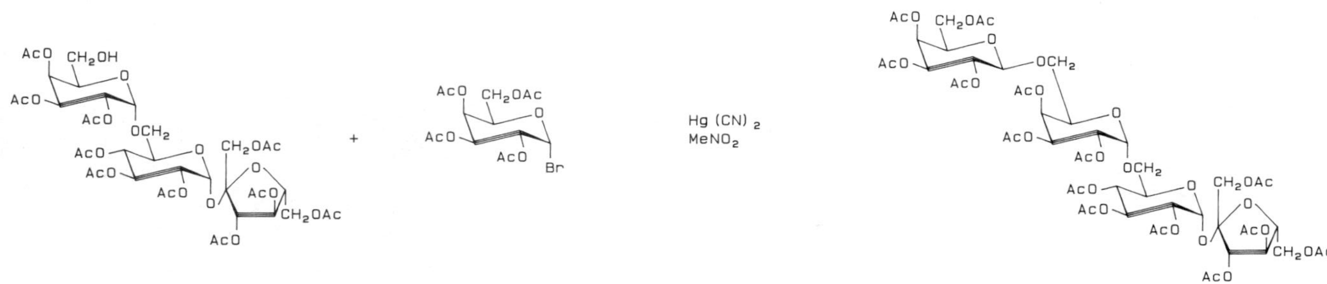

54%; m.p. 89–91 °C; $[\alpha]_D$ +69.4 (chloroform); Ref.: 150

PENTASACCHARIDES

Having monosaccharide units at the reducing end in the following sequence:

D-Xyl
D-Glc
D-GlcN (D-GlcNAc, D-GlcNSO$_3$)
D-Man
L-Rha
D-Gal

α-D-Xylp-(1→4)-β-D-Xylp-(1→4)-β-D-Xylp-(1→4)-β-D-Xylp-(1→4)-D-Xyl

β-p-(1→OMe) acetate; m.p. 240–241 °C; $[\alpha]_D$ −54.5 (chloroform); C-13; Ref.: 2

β-acetate; colourless foam; $[\alpha]_D$ −42.5 (chloroform); C-13; Ref.: 3

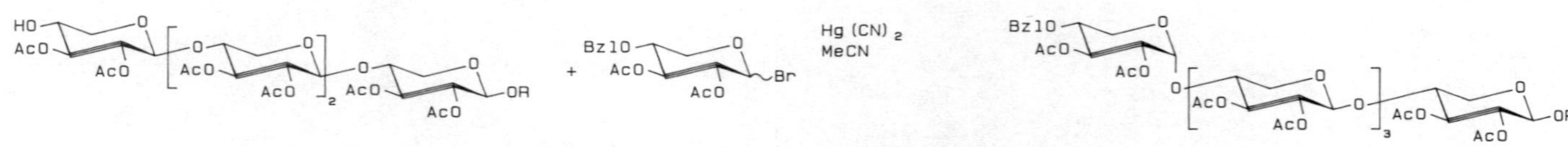

R=Me; 27.5%; amorphous solid
$[\alpha]_D$ −84.5 (chloroform); C-13; Ref.: 2

R=Ac; 28%; colourless foam
$[\alpha]_D$ −46.5 (chloroform); C-13; Ref.: 3

β-D-Xylp-(1→4)-β-D-Xylp-(1→4)-β-D-Xylp-(1→4)-β-D-Xylp-(1→4)-D-Xyl

m.p. 231–233 °C; $[\alpha]_D$ −70.1 (water); C-13; Ref.: 3

β-acetate; m.p. 251–254 °C; $[\alpha]_D$ −100.4 (chloroform); C-13; Ref.: 3

β-p-(1→OMe); m.p. 234–236 °C; $[\alpha]_D$ −90 (water); C-13; Ref.: 2

β-p-(1→OMe) acetate; m.p. 268–270 °C; $[\alpha]_D$ −115 (chloroform); C-13; Ref.: 2

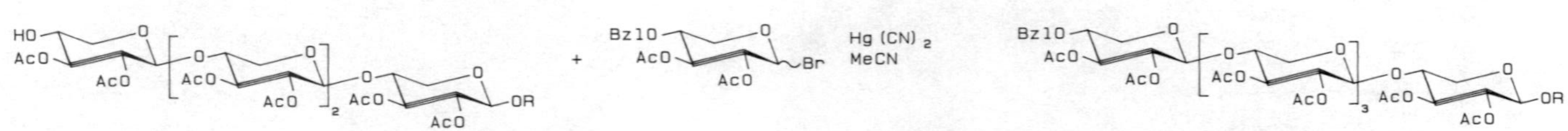

R=Me; 42%; m.p. 236–239 °C;
$[\alpha]_D$ −110 (chloroform); C-13; Ref.: 2

R=Ac; 59%; m.p. 225–227 °C;
$[\alpha]_D$ −98.7 (chloroform); C-13; Ref.: 3

β-D-Xylp-(1→3)
 ⟩β-D-Xylp-(1→4)
β-D-Xylp-(1→4)
 ⟩D-Xyl
4-O-Me-α-D-GlcpA-(1→2)

β-p-(1→OMe); glassy solid; $[\alpha]_D$ -24.6 (water); C-13; Ref.: 151

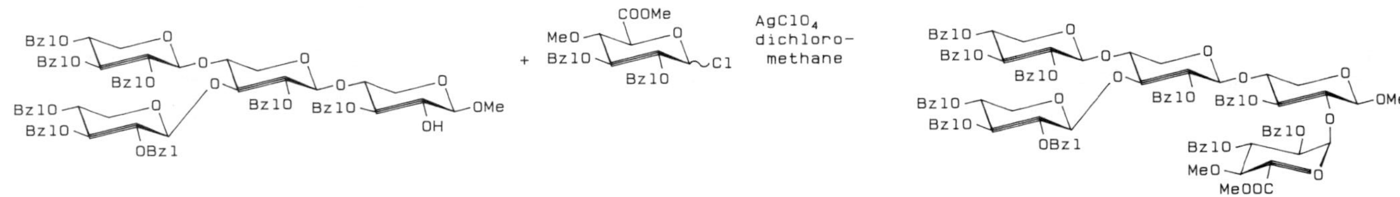

57.2%; amorphous solid;
$[\alpha]_D$ -10 (chloroform); Ref.: 151

β-D-Xylp-(1→3)
 ⟩β-D-Xylp-(1→4)
β-D-Xylp-(1→4)
 ⟩D-Xyl
4-O-Me-β-D-GlcpA-(1→2)

β-p-(1→OMe); amorphous; $[\alpha]_D$ -82 (water); C-13; Ref.: 151

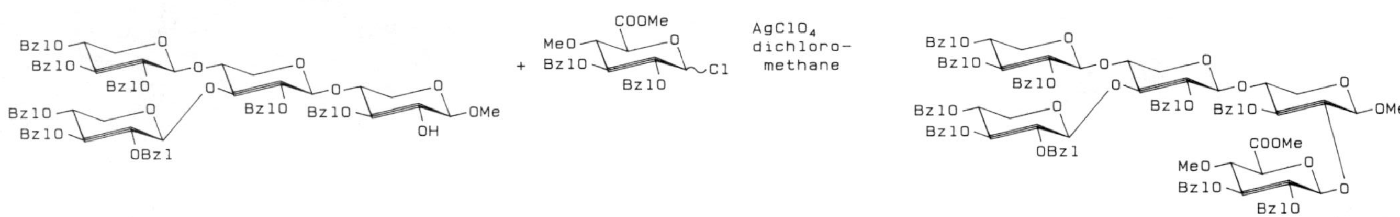

21%; amorphous solid
$[\alpha]_D$ -29 (chloroform); Ref.: 151

α-D-Glcp-(1→6)-α-D-Glcp-(1→6)-α-D-Glcp-(1→6)-α-D-Glcp-(1→6)-D-Glc

α-p-(1→OMe); m.p. 161–163 °C; $[\alpha]_D$ +182 (water); C-13; Ref.: 12

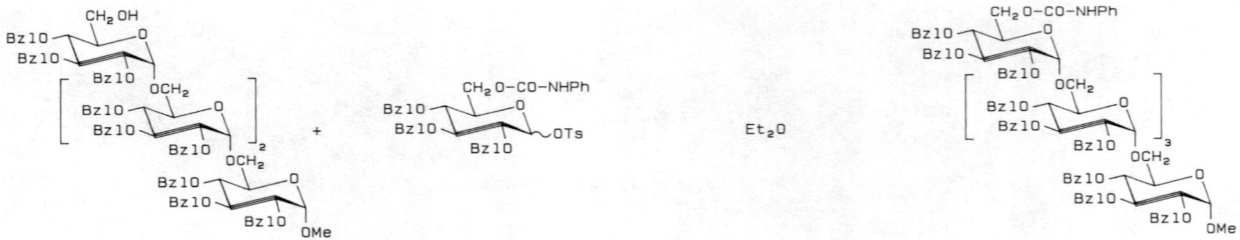

84%; $[\alpha]_D$ +81.0 (chloroform); Ref.: 11, 12

β-D-Glcp-(1→2)-β-D-Glcp-(1→2)-β-D-Glcp-(1→2)-β-D-Glcp-(1→2)-D-Glc

C-13; Ref.: 14

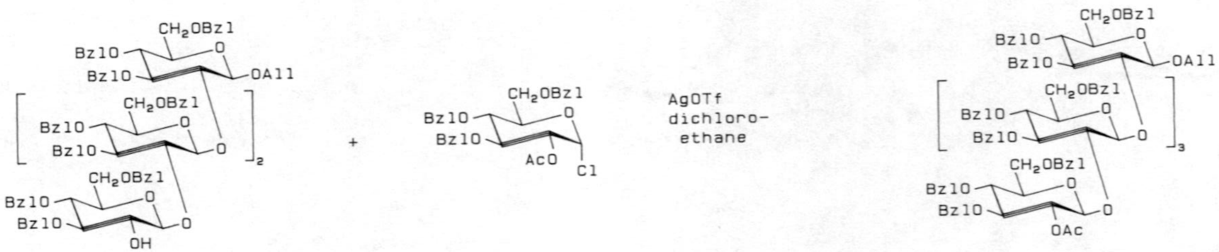

83%; $[\alpha]_D$ −3.18 (chloroform); C-13; Ref.: 14

β-D-Glcp-(1→4)-β-D-Glcp-(1→3)-β-D-Glcp-(1→4)-β-D-Glcp-(1→4)-D-Glc

β-p-(1→OMe); $[\alpha]_D$ −9.5 (water); C-13; Ref.: 152

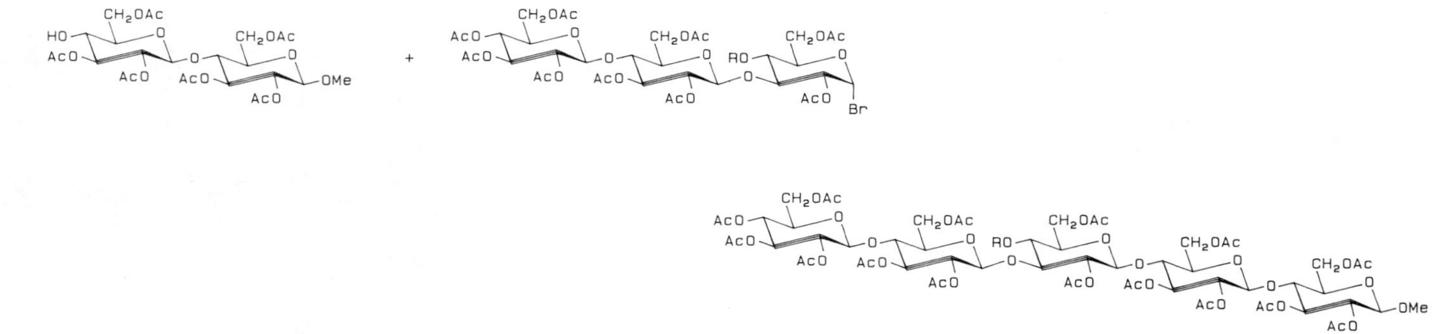

R=Ac; 39%; $[\alpha]_D$ −8.0; C-13; Ref.: 152

R=Bzl; 39%; $[\alpha]_D$ −8.4; C-13; Ref.: 152

β-D-Glcp-(1→4)-β-D-Glcp-(1→4)-β-D-Glcp-(1→4)-β-D-Glcp-(1→4)-D-Glc

α-p-(1→OR); R= 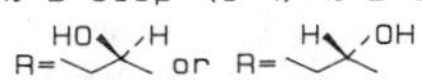 or R=

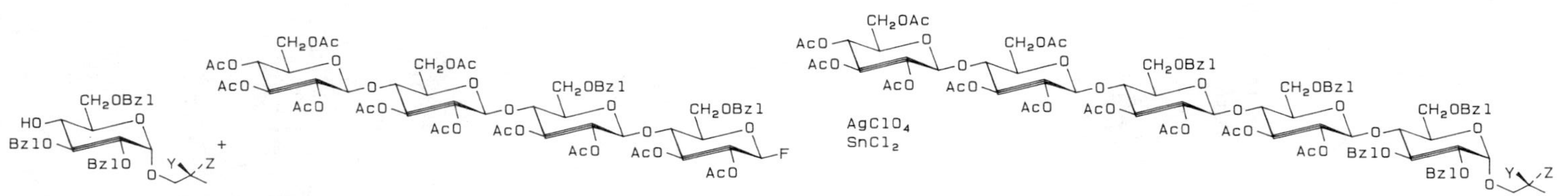

Y=OSiPh$_2$tBu; Z=H
Y=H; Z=OSiPh$_2$tBu
72%; Ref.: 19

β-D-Glcp-(1→6)-β-D-Glcp-(1→6)-β-D-Glcp-(1→6)-β-D-Glcp-(1→6)-D-Glc

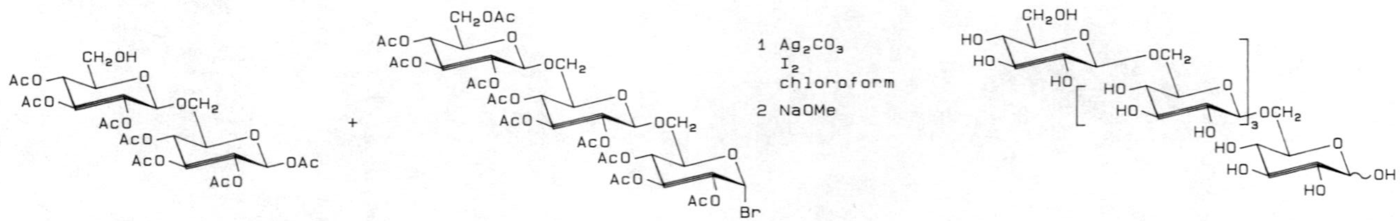

31%; Ref.: 22

α-D-GalpNAc-(1→3)-β-D-GalpNAc-(1→3)-α-D-Galp-(1→4)-β-D-Galp-(1→4)-D-Glc

$[\alpha]_D$ +134 (water); Ref.: 153

syrup; $[\alpha]_D$ +134 (water); Ref.: 154, 155

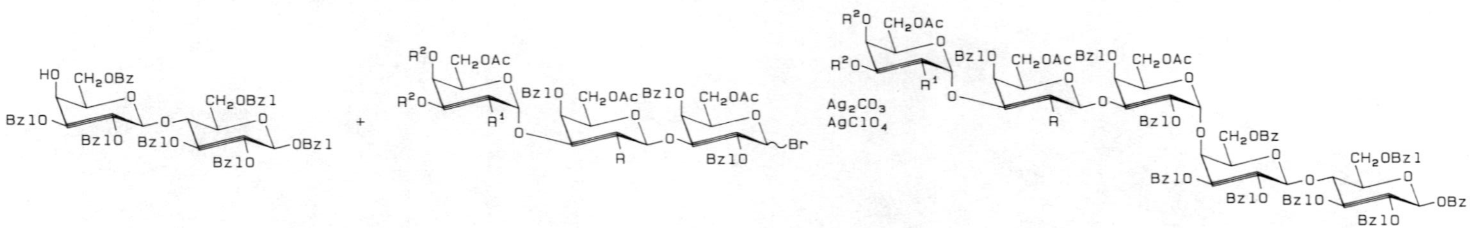

R=NPhth; $R^1=N_3$; R^2=Bz
α-Br; 28%; $[\alpha]_D$ +74.2; Ref.: 153

28%; $[\alpha]_D$ +74.1 (acetone); Ref.: 155

$R=R^1=N_3$; R^2=Bzl
38%; $[\alpha]_D$ +19.7 (chloroform); Ref.: 154

$R=R^1$=NHAc; R^2=Bzl
13%; $[\alpha]_D$ +14.8 (chloroform); Ref.: 154

β-D-Galp-(1→3)-β-D-GalpNAc-(1→4) \> β-D-Galp-(1→4)-D-Glc
α-Neu5Ac-(2→3) /

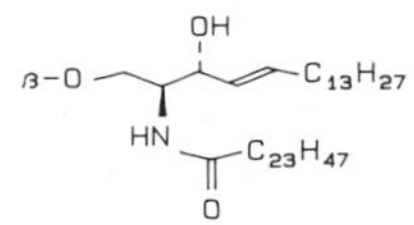

glycoside; $[\alpha]_D$ +7.3 (Py); Ref.: 49

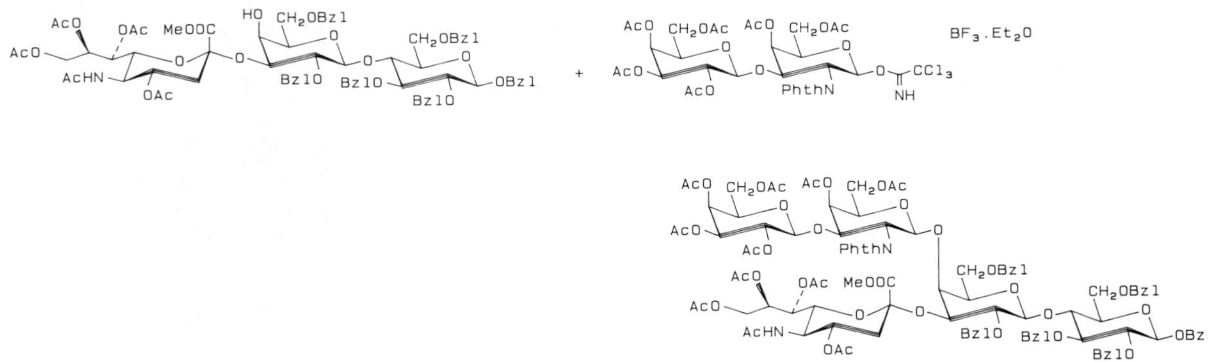

40%; $[\alpha]_D$ +6.6 (chloroform); C-13; Ref.: 49

α-D-Manp-(1→3)
α-D-Manp-(1→6) 〉β-D-Manp-(1→4)-β-D-GlcpNAc-(1→4)-D-GlcNAc

syrup; $[\alpha]_D$ +26.4 (water); Ref.: 156

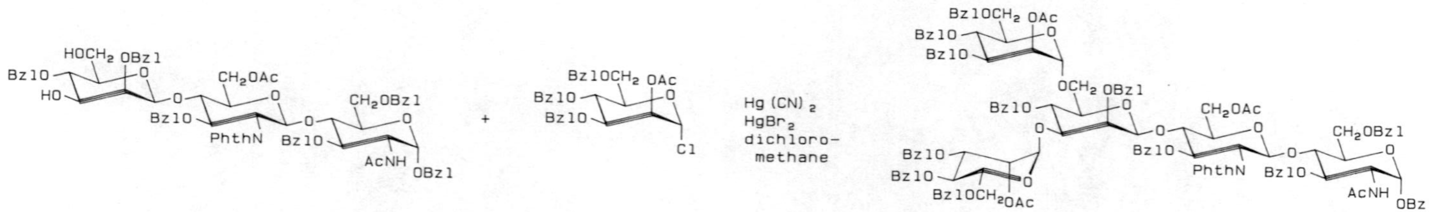

44.5%; syrup; $[\alpha]_D$ +38.8 (chloroform); C-13; Ref.: 156

β-D-Galp-(1→4)-β-D-GlcpNAc-(1→2)-α-D-Manp-(1→3)-β-D-Manp-(1→4)-D-GlcNAc

syrup; $[\alpha]_D$ +2.5 (water); Ref.: 157

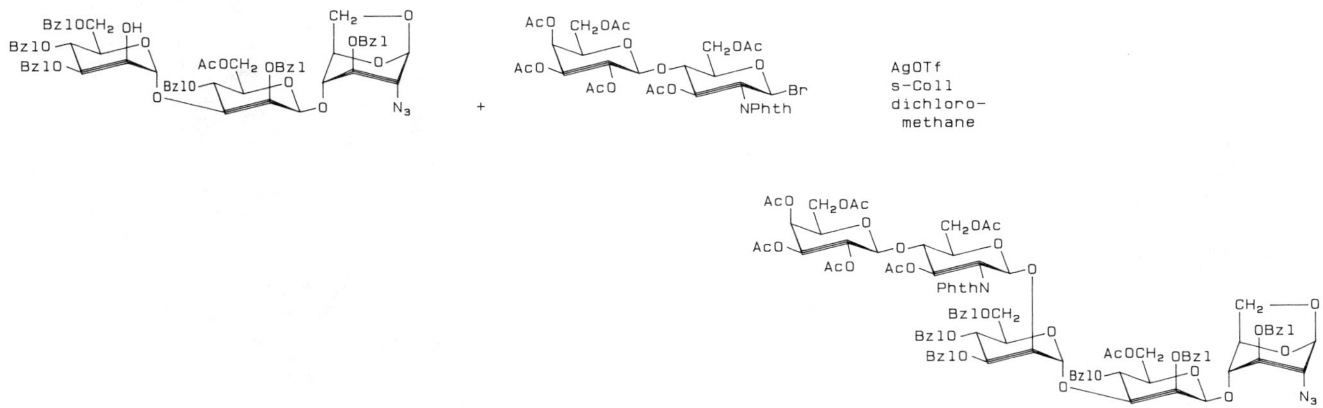

93%; syrup; $[\alpha]_D$ −13.6 (chloroform); C-13; Ref.: 157

β-D-Galp-(1→4)-β-D-GlcpNAc-(1→2)-α-D-Manp-(1→6)-β-D-Manp-(1→4)-D-GlcNAc

syrup; $[\alpha]_D$ +6.0 (water); Ref.: 157, 158

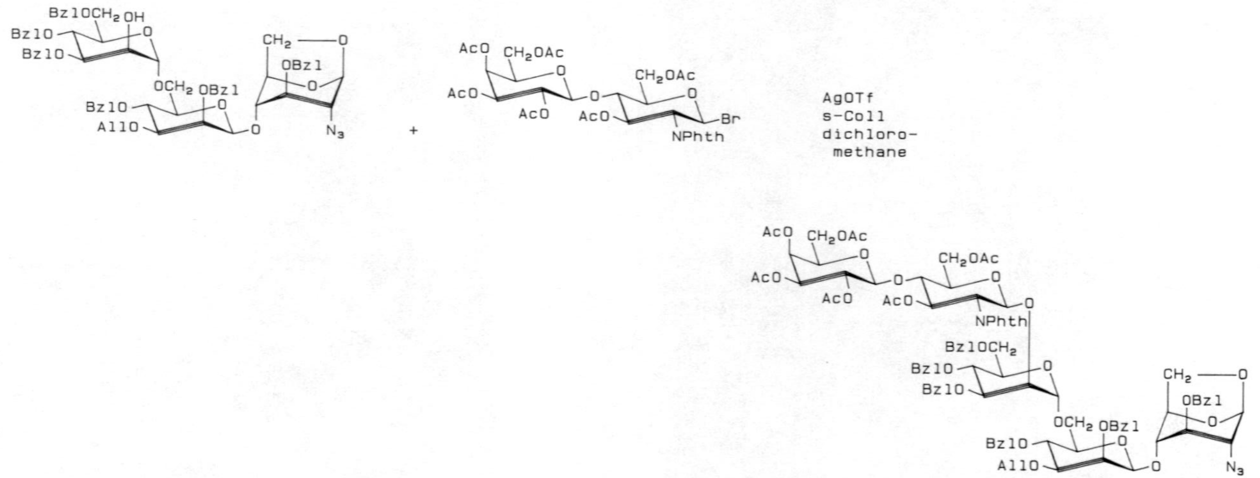

75%; syrup; $[\alpha]_D$ −9.7 (chloroform); C-13; Ref.: 157, 158

β-D-GlcpNAc-(1→2)-α-D-Manp-(1→6)
α-D-Manp-(1→3) 〉β-D-Manp-(1→4)-D-GlcNAc

$[\alpha]_D$ -4.3 (water); Ref.: 159

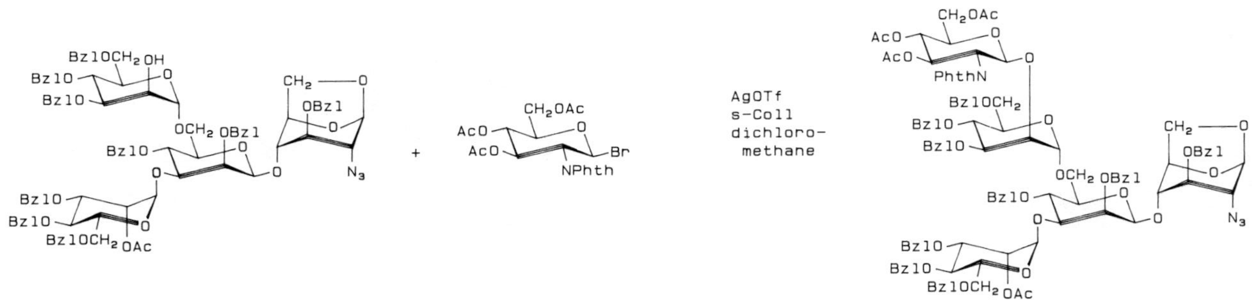

73%; syrup; $[\alpha]_D$ +3 (chloroform); Ref.: 159

β-D-GlcpNAc-(1→2)-α-D-Manp-(1→3) / α-D-Manp-(1→6) > β-D-Manp-(1→4)-D-GlcNAc

syrup; $[\alpha]_D$ -7.7 (water); Ref.: 159

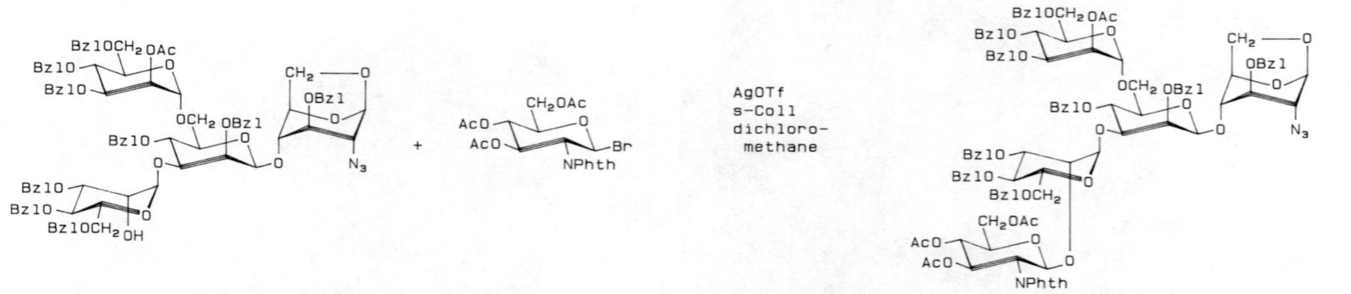

61%; syrup; $[\alpha]_D$ +8.6 (chloroform); Ref.: 159

α-D-Manp-(1→3)

β-D-GlcpNAc-(1→4) → β-D-Manp-(1→4)-D-GlcNAc

α-D-Manp-(1→6)

syrup; $[\alpha]_D$ +42.4 (water); Ref.: 160

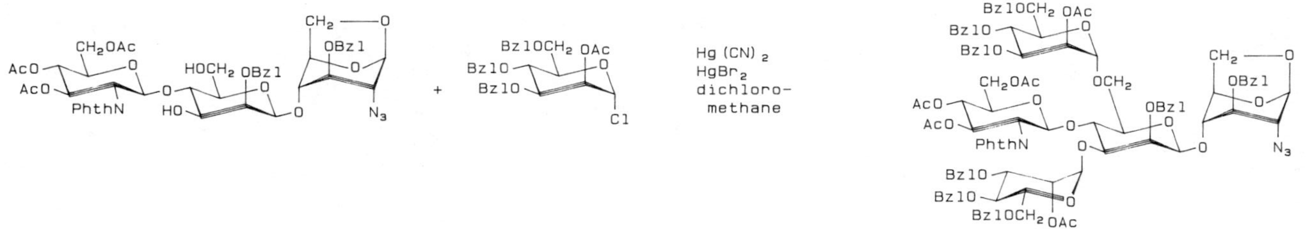

68%; amorphous
$[\alpha]_D$ +27.5 (chloroform); C-13; Ref.: 160

α-D-GlcpNSO$_3$(6-OSO$_3$)-(1→4)-β-D-GlcpA-(1→4)-α-D-GlcpNSO$_3$(3, 6-OSO$_3$)-(1→4)-α-L-IdopA(2-OSO$_3$)-(1→4)-D-GlcNSO$_3$

α-p-(1→OMe); C-13; Ref.: 67

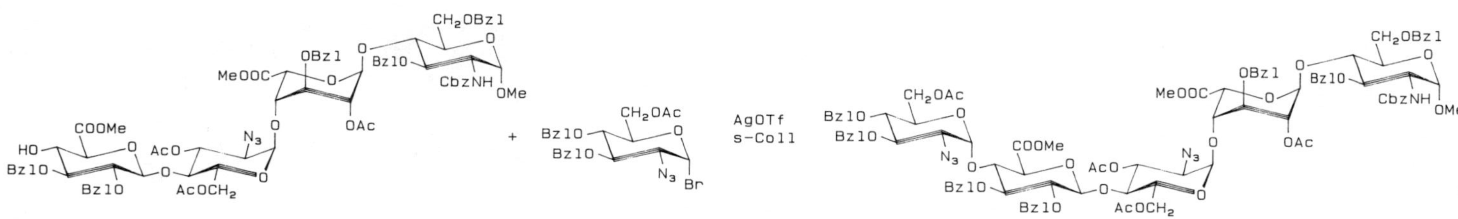

78%; Ref.: 67

α-D-GlcpNSO$_3$(6-OSO$_3$)-(1→4)-β-D-GlcpA-(1→4)-α-D-GlcpNSO$_3$(3,6-OSO$_3$)-(1→4)-α-L-IdopA(2-OSO$_3$)-(1→4)-D-GlcNSO$_3$(6-OSO$_3$)

sodium salt; amorphous white powder; $[\alpha]_D$ +42 (water); C-13; Ref.: 68, 69

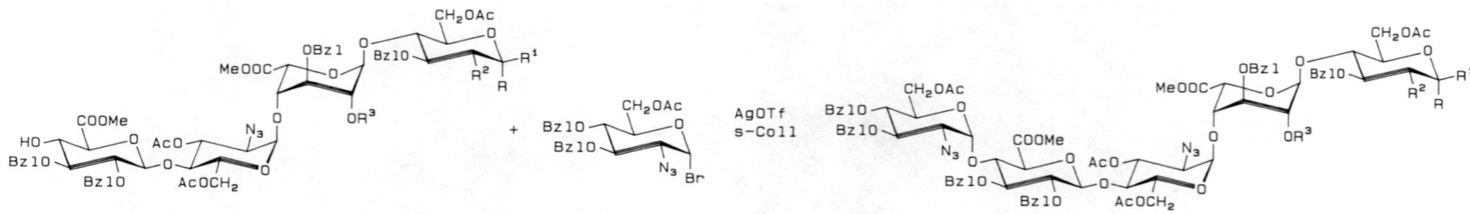

R=OBzl; R^1=H; R^2=NHCbz; R^3=Ac
70-89%; $[\alpha]_D$ +65 (chloroform); Ref.: 68, 69

R=H; R^1=OBzl; R^2=N$_3$; R^3=Ac
96%; $[\alpha]_D$ 27.5 (chloroform); C-13; Ref.: 70

R=H; R^1=OBzl; R^2=N$_3$; R^3=Bz
70%; $[\alpha]_D$ +21; Ref.: 71

β-D-GlcpN-(1→4)-β-D-GlcpA-(1→4)-α-D-GlcpN-(1→4)-α-L-IdopA-(1→4)-D-GlcN

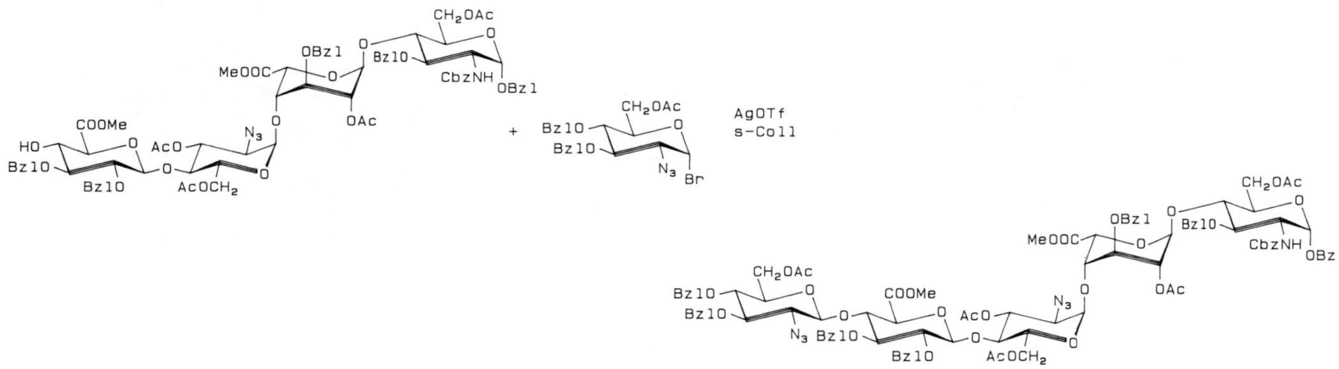

14%; $[\alpha]_D$ +46 (chloroform); Ref.: 69

α-D-Manp-(1→3)-α-D-Manp-(1→2)-α-D-Manp-(1→2)-α-D-Manp-(1→2)-D-Man

$[\alpha]_D$ +27.2 (water); C-13; Ref.: 98

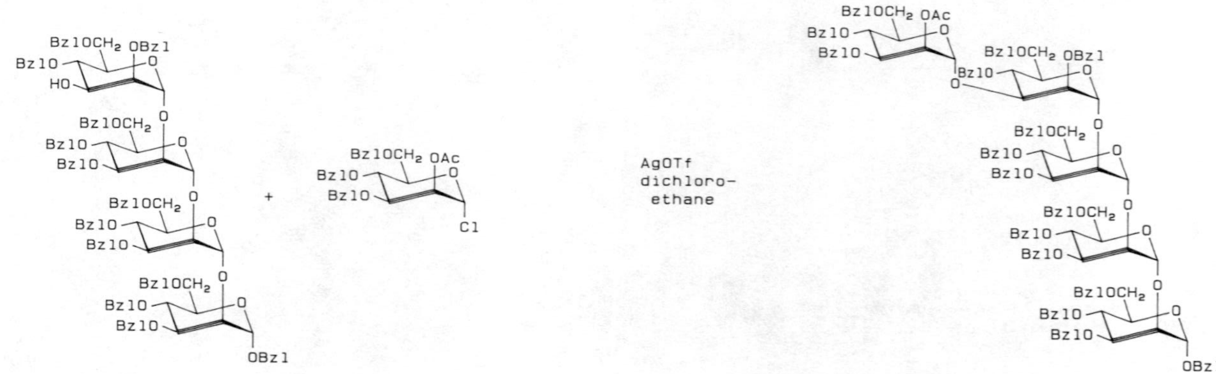

58%; Ref.: 98

α-D-Manp-(1→3)-α-D-Manp-(1→3)-α-D-Manp-(1→3)-α-D-Manp-(1→3)-D-Man

amorphous solid; $[\alpha]_D$ +25.6 (water); C-13; Ref.: 99

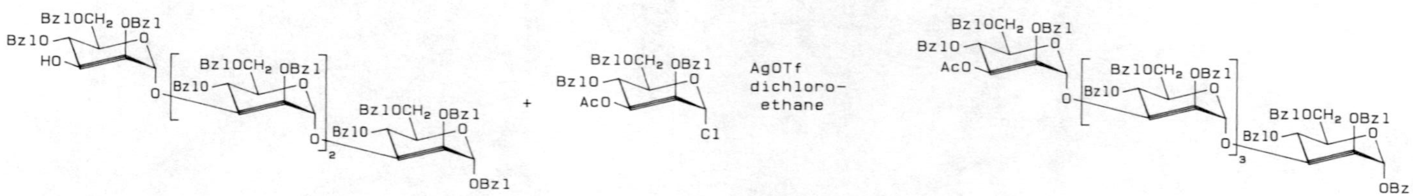

96%; $[\alpha]_D$ +12.2 (chloroform); Ref.: 99

β-D-Galp-(1→4)-β-D-GlcpNAc-(1→2)
β-D-Galp-(1→4)-β-D-GlcpNAc-(1→4) >D-Man

amorphous powder; $[\alpha]_D$ −13 (water); C-13; Ref.: 161, 162

α-p-(1→O-p-Trifluoroacetamidophenyl); amorphous powder; $[\alpha]_{578}$ +10 (water); C-13; Ref.: 163

α-p-(1→OPr); $[\alpha]_D$ 0 (water); C-13; Ref.: 164

α-p-(1→OMCO); $[\alpha]_D$ +1 (water); Ref.: 164

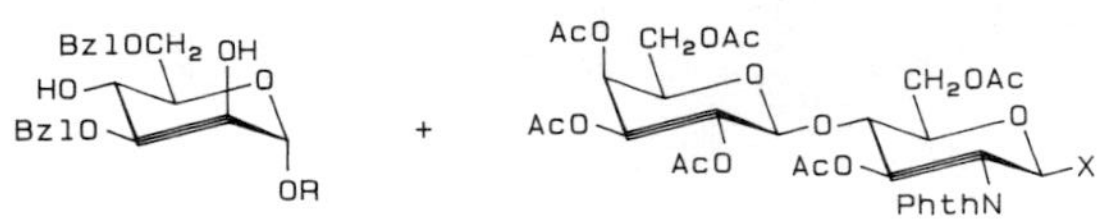

I AgOTf
s-Coll
dichloro-
methane
X=Br

II AgOTf
s-Coll
dichloro-
ethane
X=Br

III $BF_3.Et_2O$
dichloro-
ethane
X=O-CNH-CCl_3

IV MeOTf
dichloro-
ethane
Et_2O
X=SMe

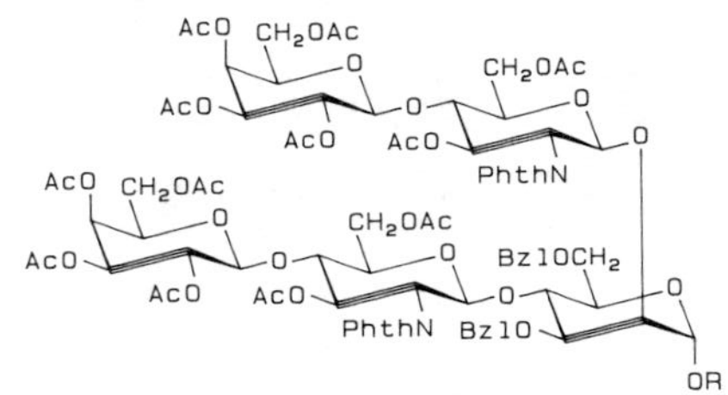

I R=Bzl; syrup; 56%
$[\alpha]_D$ +12 (chloroform); C-13; Ref.: 161, 162

R=PNP; syrup; 41%
$[\alpha]_{578}$ +47 (chloroform); C-13; Ref.: 163

II R=All; 57.3%
$[\alpha]_D$ +15.2 (chloroform); C-13; Ref.: 164

III R=All; 73.3%; Ref.: 164

IV R=All; 36%; Ref.: 164

β-D-Galp-(1→4)-β-D-GlcpNAc-(1→2)
β-D-Galp-(1→4)-β-D-GlcpNAc-(1→6) >D-Man

amorphous powder; $[\alpha]_D$ −21 (water); C-13; Ref.: 165

α-p-(1-O-p-Trifluoroacetamidophenyl); amorphous powder; $[\alpha]_D$ +2 (water); C-13; Ref.: 163

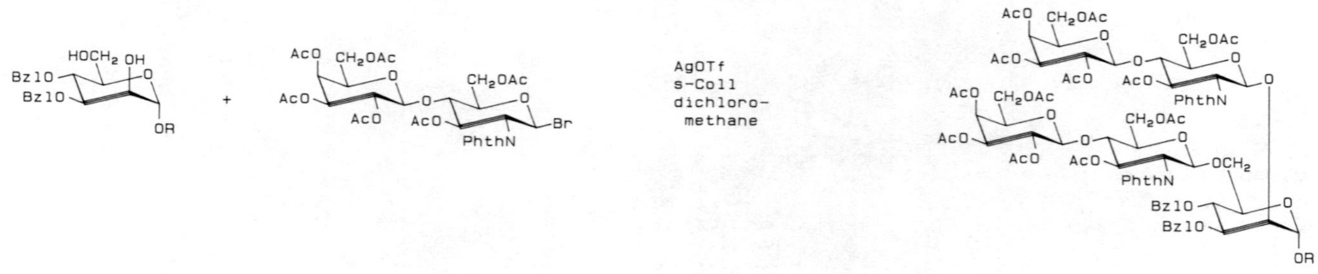

I R=Bzl; 52%; m.p. 116–119 °C;
$[\alpha]_D$ −8 (chloroform); C-13; Ref.: 165

R=PNP; 78%; amorphous powder
$[\alpha]_D$ +58 (chloroform); C-13; Ref.: 163

α-D-Manp-(1→2)-α-D-Manp-(1→2)
α-D-Manp-(1→2)-α-D-Manp-(1→4) > D-Man

$[\alpha]_D$ +85.2 (water); C-13; Ref.: 166

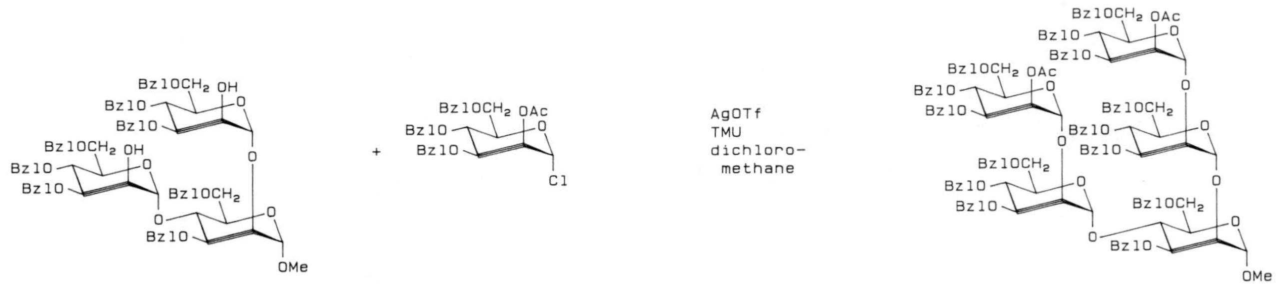

87.8%; $[\alpha]_D$ +19.2 (chloroform); C-13; Ref.: 166

α-D-Manp-(1→2)-α-D-Manp-(1→2)
α-D-Manp-(1→2)-α-D-Manp-(1→6) 〉D-Man

amorphous; $[\alpha]_D$ +71.9 (water); C-13; Ref.: 166

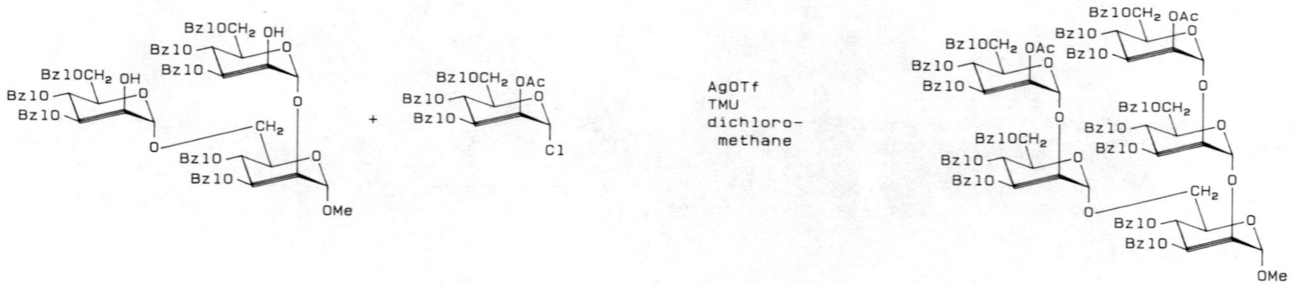

56.7%; $[\alpha]_D$ +33.9 (chloroform); Ref.: 166

β-D-Galp-(1→4)-β-D-GlcpNAc-(1→2)-α-D-Manp-(1→3)
α-D-Manp-(1→6) 〉D-Man

$[\alpha]_D$ +4.9 (water); C-13; Ref.: 167

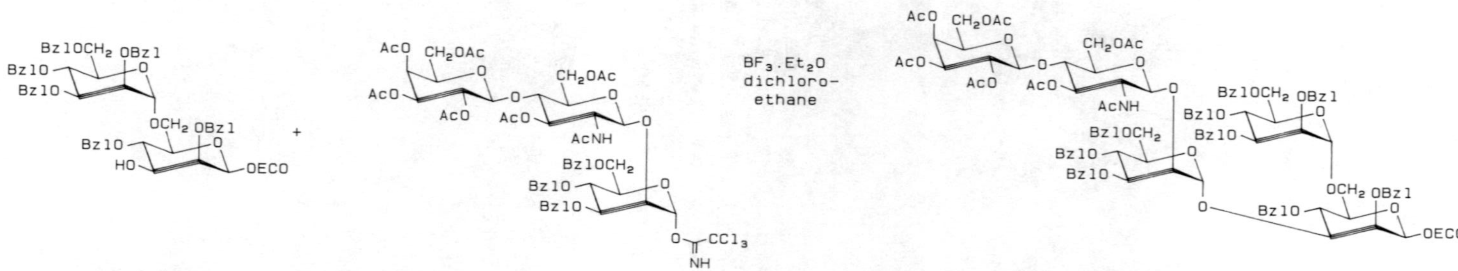

45%; $[\alpha]_D$ -8.7 (chloroform); C-13; Ref.: 167

β-D-Galp-(1→4)-β-D-GlcpNAc-(1→4)-α-D-Manp-(1→3)
α-D-Manp-(1→6) 〉D-Man

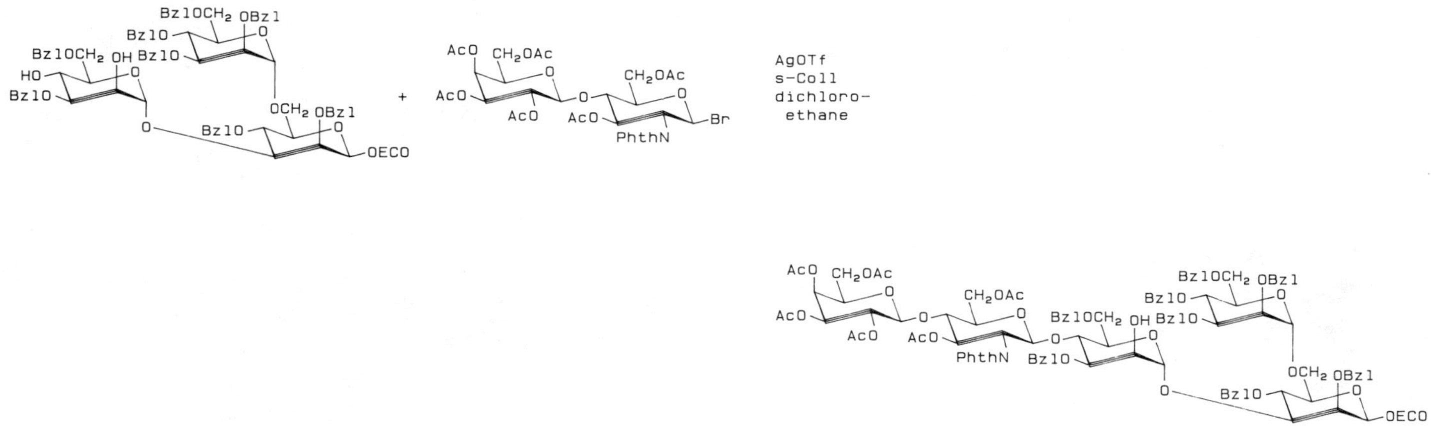

32.6%; $[\alpha]_D$ +14 (chloroform); Ref.: 168

33%; $[\alpha]_D$ +14.4 (chloroform); Ref.: 169

β-D-GlcpNAc-(1→2)-α-D-Manp-(1→3) \
β-D-GlcpNAc-(1→2)-α-D-Manp-(1→6) / D-Man

α-p-(1-OMe); $[\alpha]_D$ +44.2 (MeOH); C-13; Ref.: 103, 170

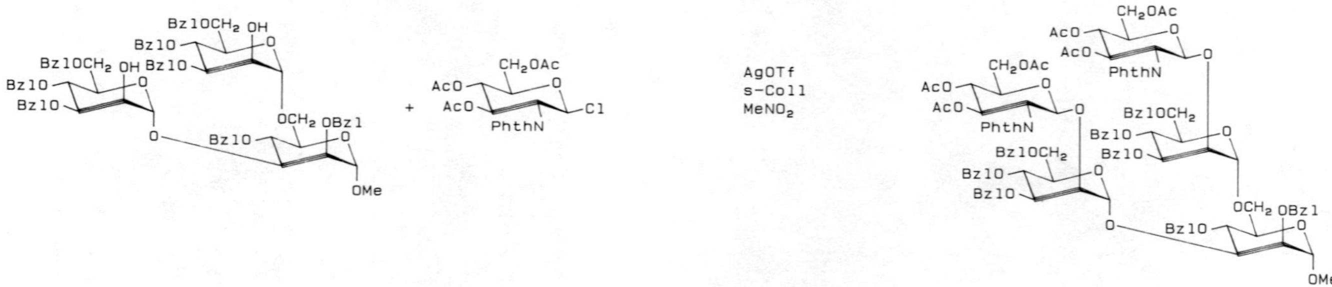

43.7%; $[\alpha]_D$ +12.0 (chloroform); Ref.: 103, 170

α-D-Manp-(1→2)-α-D-Manp-(1→3) \
α-D-Manp-(1→2)-α-D-Manp-(1→6) / D-Man

α-p-(1-OMe); $[\alpha]_D$ +89.6 (MeOH); C-13; Ref.: 103, 170

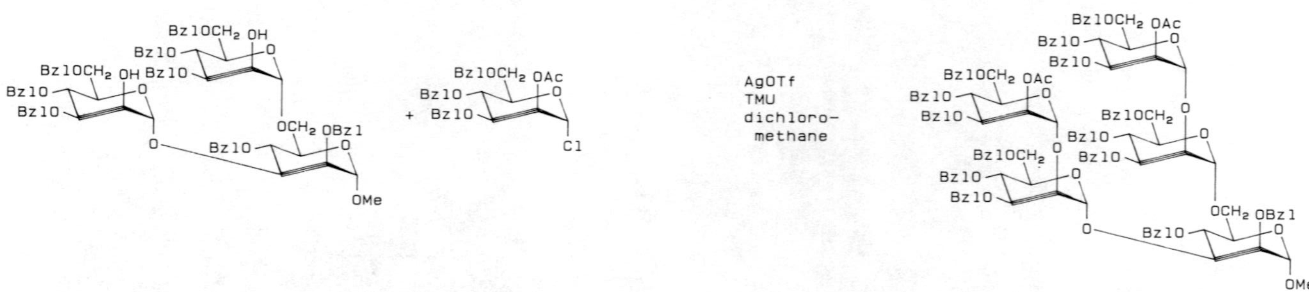

78.8%; $[\alpha]_D$ +34.7 (chloroform); Ref.: 103, 170

α-D-Manp-(1→2) α-D-Manp-(1→6) α-D-Manp-(1→6) D-Man

α-D-Manp-(1→6) (branch at α-D-Manp-(1→6)); α-D-Manp-(1→3) (branch at D-Man)

α-p-(1→OMe); amorphous powder; $[\alpha]_D$ +91.7 (water); C-13; Ref.: 171

74.8%; $[\alpha]_D$ +40.3 (chloroform); C-13; Ref.: 171

α-D-Manp-(1→3) ⟩ α-D-Manp-(1→3) ⟩ D-Man
α-D-Manp-(1→6) ⟩ α-D-Manp-(1→6) ⟩

α-p-(1-OMe); amorphous; $[\alpha]_D$ +98.3 (water); C-13; Ref.: 172

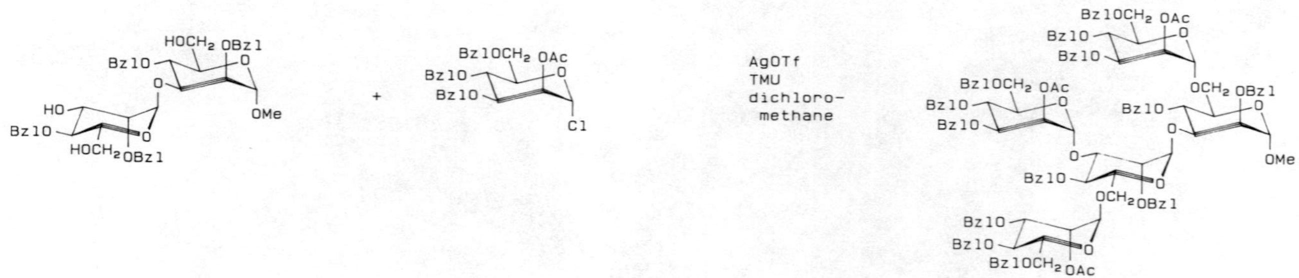

76.1%; $[\alpha]_D$ +57.1 (chloroform); C-13; Ref.: 172

α-D-Manp-(1→3)
α-D-Manp-(1→6) > α-D-Manp-(1→6)
α-D-Manp-(1→3) > D-Man

α-p-(1→OMe); $[\alpha]_D$ +108.1 (water); +115 (MeOH); Ref.: 173

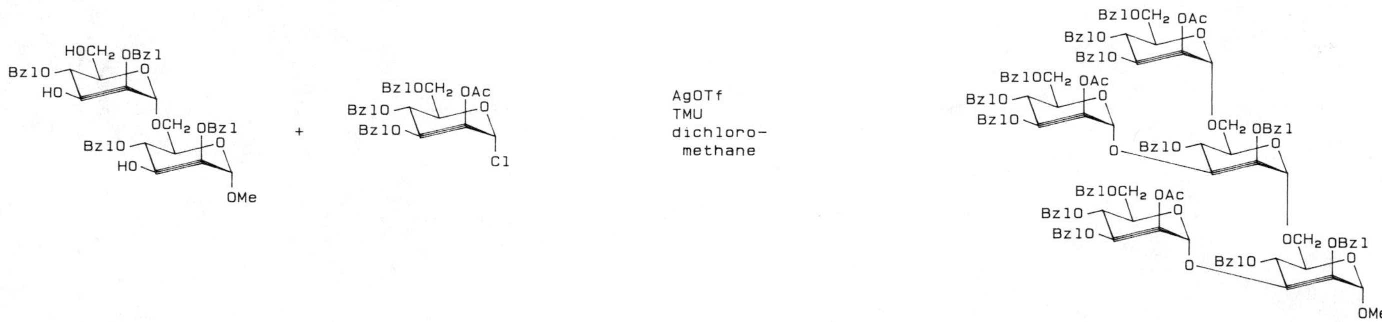

73%; $[\alpha]_D$ +45.2 (chloroform); Ref.: 173

β-D-GlcpNAc-(1→2)
α-D-Glcp-(1→3) 〉α-L-Rhap-(1→2)
α-D-Glcp-(1→3) 〉L-Rha

α-p-(1→OMe); white amorphous powder; $[\alpha]_D$ +68 (MeOH), +71 (water); C-13; Ref.: 126, 127

α-p-(1→OMe); $[\alpha]_D$ +46.7 (MeOH); C-13; Ref.: 128

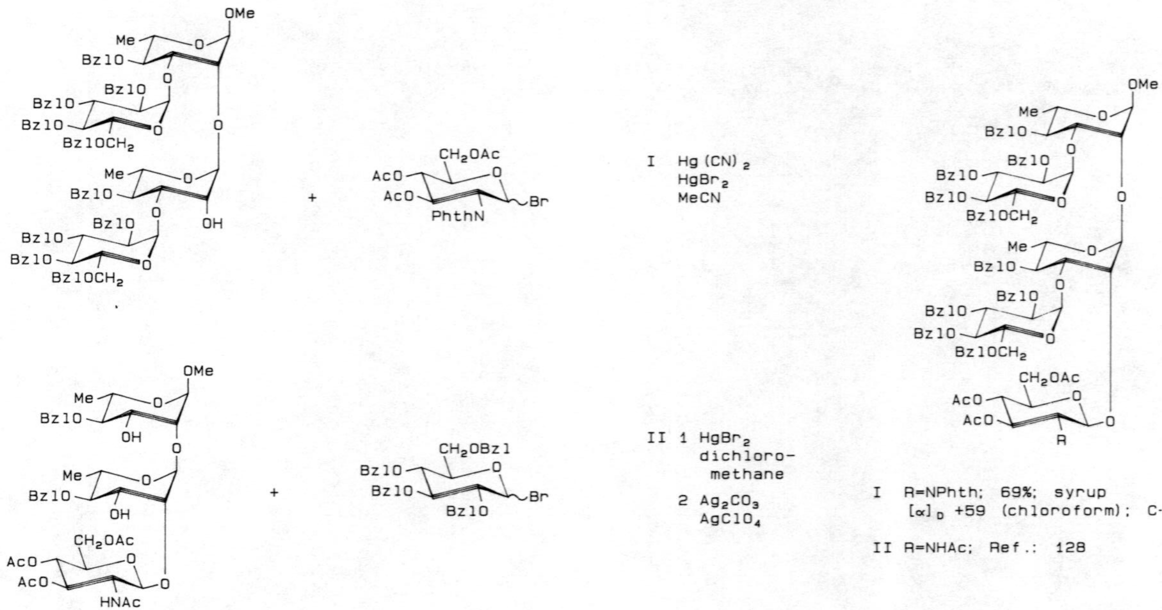

I R=NPhth; 69%; syrup
$[\alpha]_D$ +59 (chloroform); C-13; Ref.: 126, 127

II R=NHAc; Ref.: 128

α-D-GalpNAc-(1→3)
α-D-GlcpNAc-(1→4) 〉 α-D-GalpNAc-(1→4)-α-D-Glcp-(1→4)-D-Gal

$[\alpha]_D$ +170 (water); Ref.: 51, 52

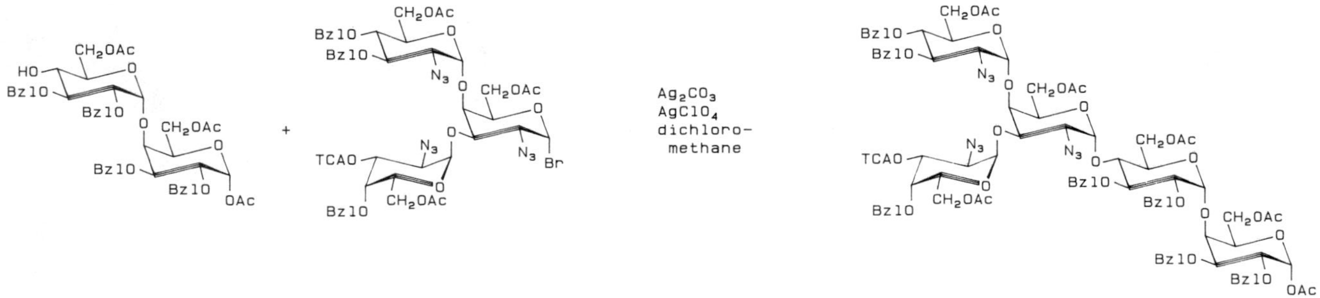

19%; syrup; $[\alpha]_D$ +102 (chloroform); Ref.: 51, 52

β-D-Galp-(1→3)-β-D-Galp-(1→3)-β-D-Galp-(1→3)-β-D-Galp-(1→3)-D-Gal

β-p-(1→OMe); amorphous, hygroscopic solid; $[\alpha]_D$ +34.6 (water); C-13; Ref.: 132

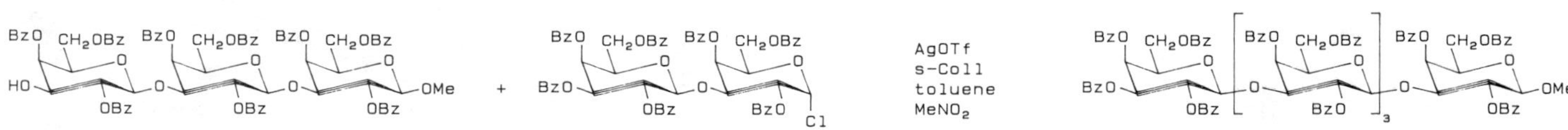

69%; amorphous solid
$[\alpha]_D$ +45 (chloroform); C-13; Ref.: 132

β-D-Galp-(1→6)-β-D-Galp-(1→6)-β-D-Galp-(1→6)-β-D-Galp-(1→6)-D-Gal

β-p-(1→OMe); m.p. 275 °C (dec. with softening at cca. 180 °C); $[\alpha]_D$ -9.7 (water); C-13; Ref.: 136

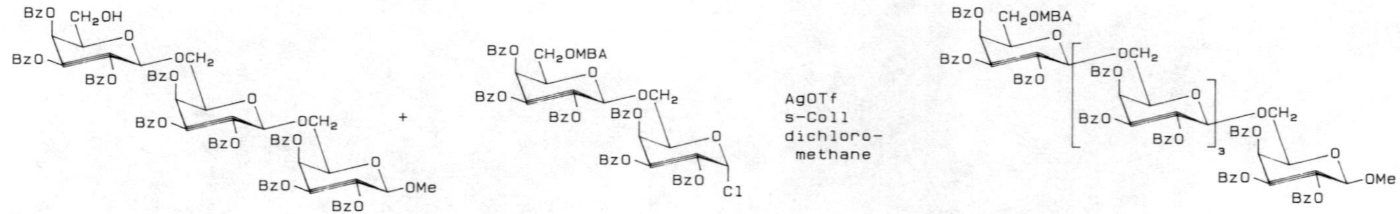

67.7%; m.p. 263-264 °C;
$[\alpha]_D$ +85.6 (chloroform); C-13; Ref.: 136, 137

α-D-Tyvp-(1→3)-β-D-Manp-(1→4)-α-L-Rhap-(1→3) ⟩ D-Gal
α-D-Glcp-(1→4) ⟩

$[\alpha]_D$ +42.5 (water); C-13; Ref.: 174

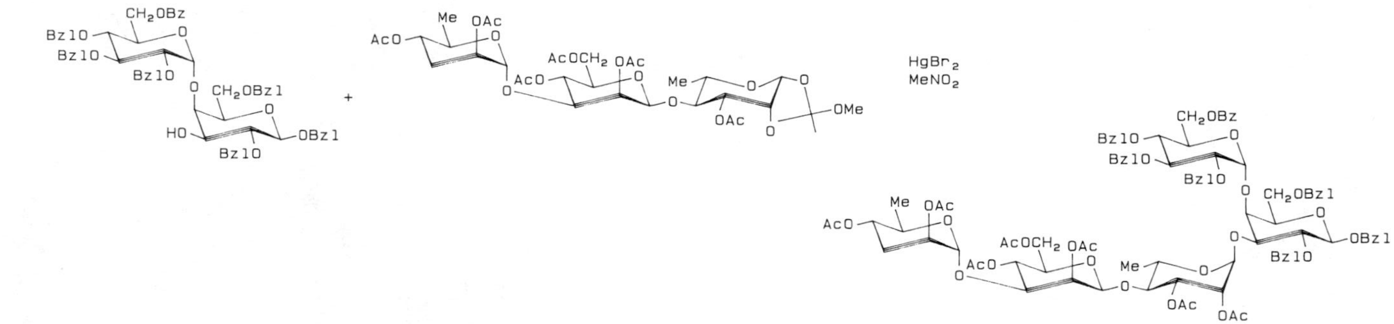

17.5%; $[\alpha]_D$ +21 (chloroform); Ref.: 174

β-D-Galp-(1→3)-β-D-GlcpNAc-(1→3) ⟩ D-Gal
β-D-Galp-(1→4)-β-D-GlcpNAc-(1→6) ⟩

m.p. 206–210 °C (dec.); $[\alpha]_D$ −6 → +1.5 (water); Ref.: 142, 143

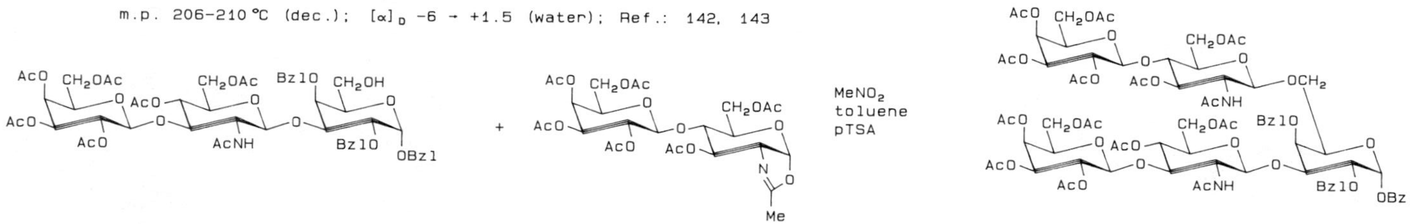

55%; foam; m.p. 125–129 °C;
$[\alpha]_D$ +1.9 (chloroform); Ref.: 142, 143

β-D-Galp-(1→4)-β-D-GlcpNAc-(1→3)
β-D-Galp-(1→4)-β-D-GlcpNAc-(1→6) 〉D-Gal

$[\alpha]_D$ +1.6 (water); Ref.: 143

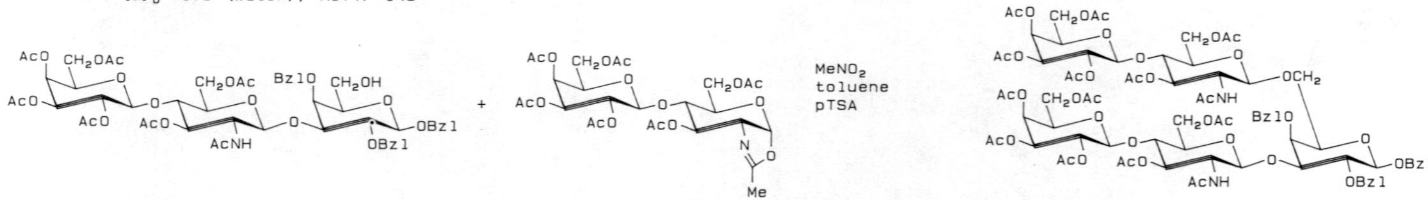

66%; m.p. 128–130 °C; $[\alpha]_D$ −11.5 (chloroform); Ref.: 143

HEXASACCHARIDES

Having monosaccharide units at the reducing end in the following sequence:

D-Xyl
D-Glc
D-GlcN (D-GlcNAc)
D-Man
D-Gal

α-D-Xylp-(1→4)-β-D-Xylp-(1→4)-β-D-Xylp-(1→4)-β-D-Xylp-(1→4)-β-D-Xylp-(1→4)-D-Xyl

β-p-(1→OMe) acetate; amorphous solid; $[\alpha]_D$ -68 (chloroform); C-13; Ref.: 2

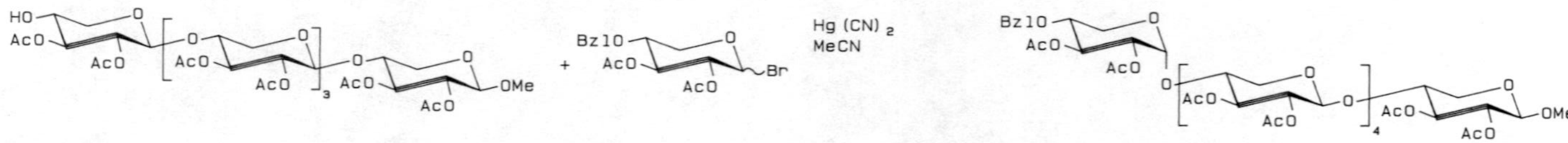

21%; amorphous solid
$[\alpha]_D$ -99 (chloroform); C-13; Ref.: 2

β-D-Xylp-(1→4)-β-D-Xylp-(1→4)-β-D-Xylp-(1→4)-β-D-Xylp-(1→4)-β-D-Xylp-(1→4)-D-Xyl

β-p-(1→OMe); m.p. 254-256 °C; $[\alpha]_D$ -92 (water); C-13; Ref.: 2

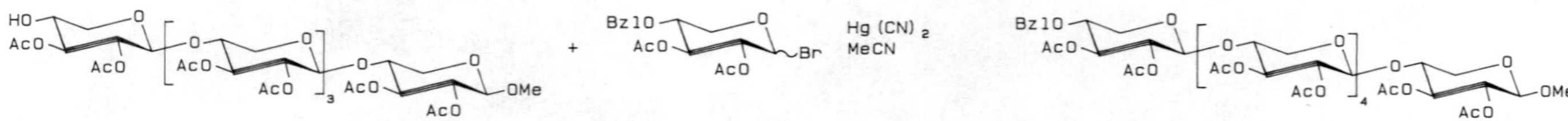

39.5%; m.p. 247-250 °C;
$[\alpha]_D$ -114 (chloroform); C-13; Ref.: 2

α-D-Glcp-(1→4)-α-D-Glcp-(1→4)-α-D-Glcp-(1→4)-α-D-Glcp-(1→4)-α-D-Glcp-(1→4)-D-Glc

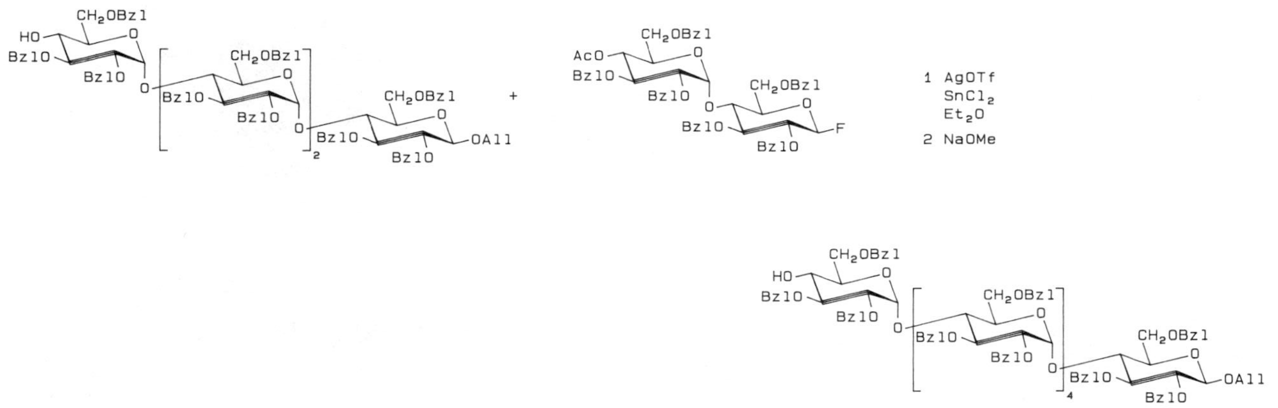

1 65%; $\alpha/\beta=2:1$

2 79%; $[\alpha]_D$ +64.7 (chloroform); C-13; Ref.: 6

α-D-Glcp-(1→4)-β-D-Glcp-(1→4)-α-D-Glcp-(1→4)-α-D-Glcp-(1→4)-α-D-Glcp-(1→4)-D-Glc

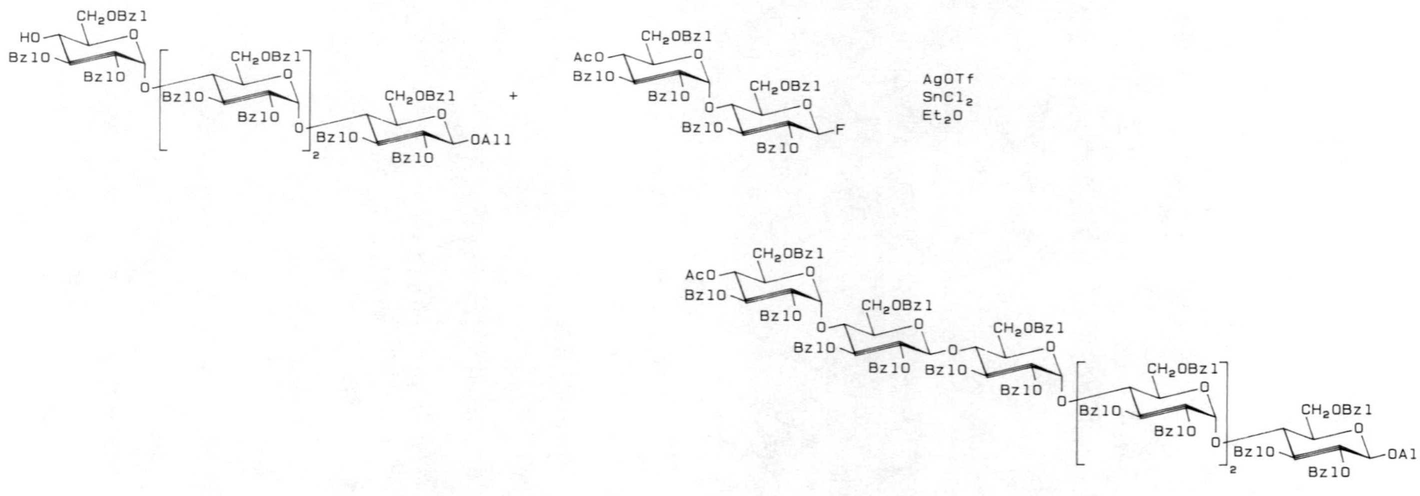

65%; α/β=2: 1; Ref.: 6

α-Cyclodextrin

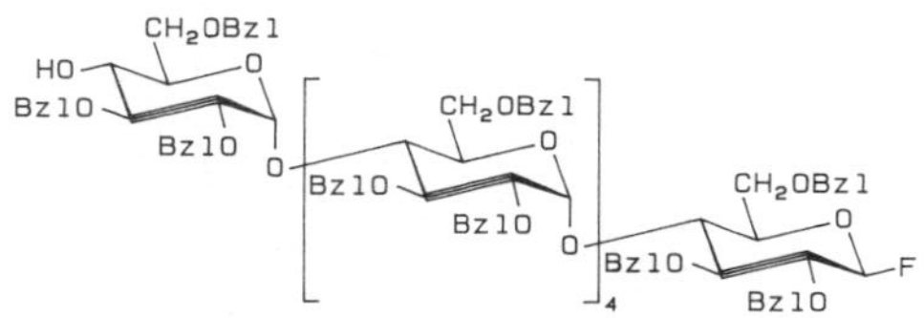

dichloro-
ethane
$SnCl_2$
AgOTf
Et_2O

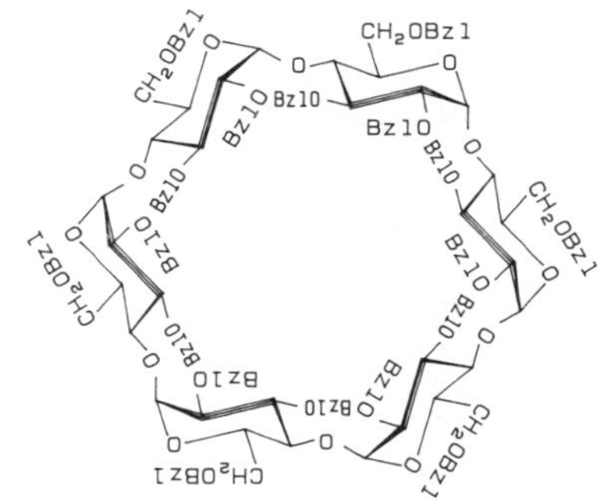

21 %
$[\alpha]_D$ +49.6 (chloroform); C-13; Ref.: 6

α-D-Glcp-(1→6)-α-D-Glcp-(1→6)-α-D-Glcp-(1→6)-α-D-Glcp-(1→6)-α-D-Glcp-(1→6)-D-Glc

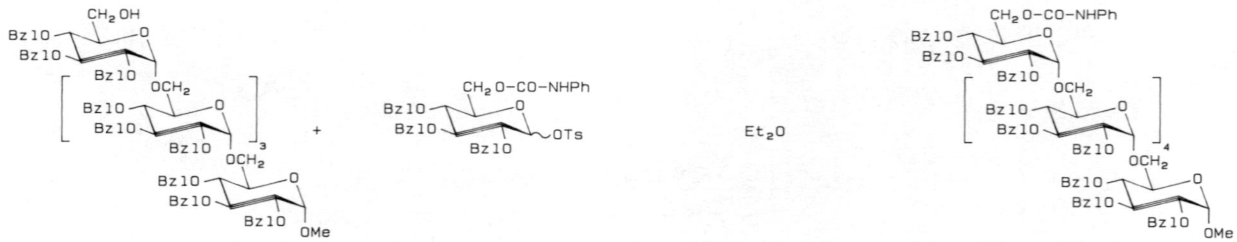

83%; $[\alpha]_D$ +85.5 (chloroform); Ref.: 11, 12

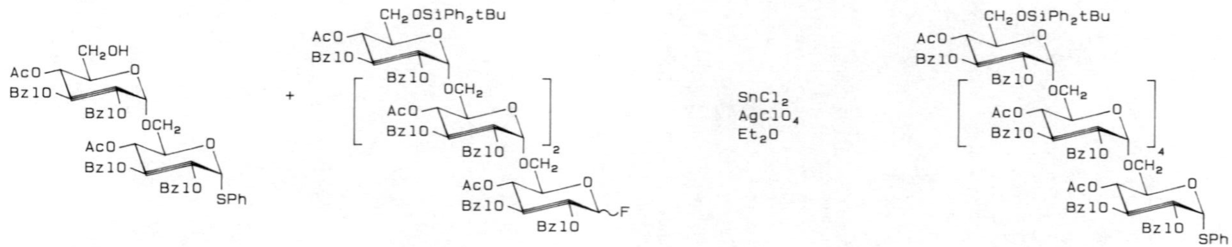

66%; $[\alpha]_D$ +66.5 (dichloromethane); Ref.: 10

β-D-Glcp-(1→6)-β-D-Glcp-(1→6)-β-D-Glcp-(1→6)-β-D-Glcp-(1→6)-β-D-Glcp-(1→6)-D-Glc

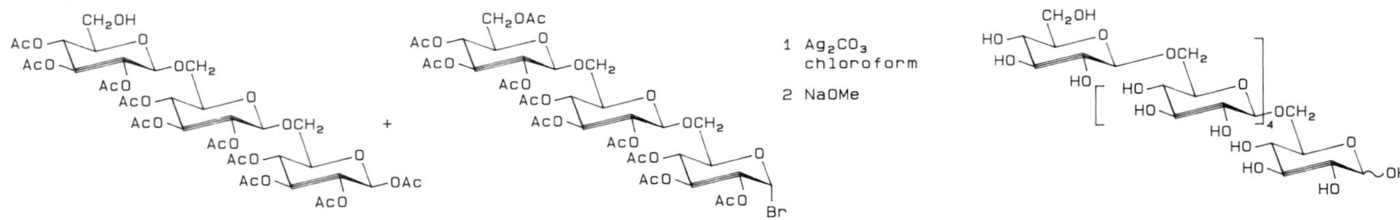

Ref.: 22

α-L-Fucp-(1→2)-β-D-Galp-(1→3)
>β-D-GlcpNAc-(1→3)-β-D-Galp-(1→4)-D-Glc
α-L-Fucp-(1→4)

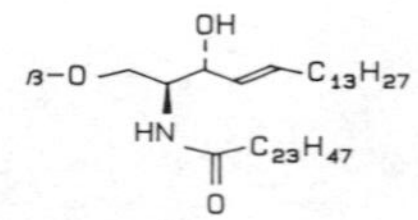

glycoside; $[\alpha]_D$ -42.1; Ref.: 33

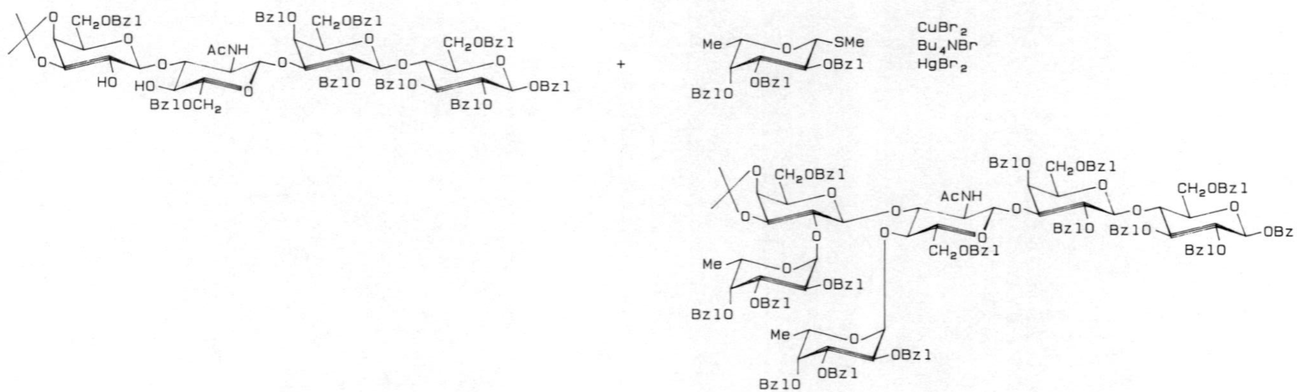

68%; $[\alpha]_D$ -37 (chloroform); Ref.: 33

β-D-Galp-(1→4)-β-D-GlcpNAc-(1→6)
β-D-Galp-(1→4)-β-D-GlcpNAc-(1→3) 〉 β-D-Galp-(1→4)-D-Glc

m.p. 223–225 °C; $[\alpha]_D$ +9.1 (water); Ref.: 40, 41

acetate; amorphous powder; $[\alpha]_D$ +12.7 (chloroform); Ref.: 39

β-p-(1→OMe); $[\alpha]_D$ −15 (water); C-13; Ref.: 39

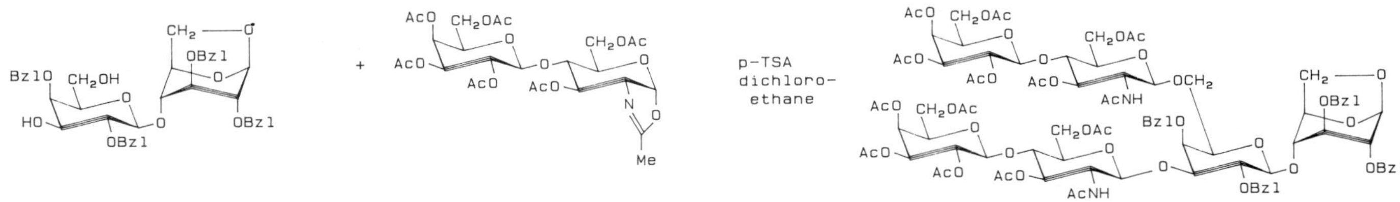

53.5%; amorphous powder;
$[\alpha]_D$ −13.8 (chloroform); Ref.: 40, 41

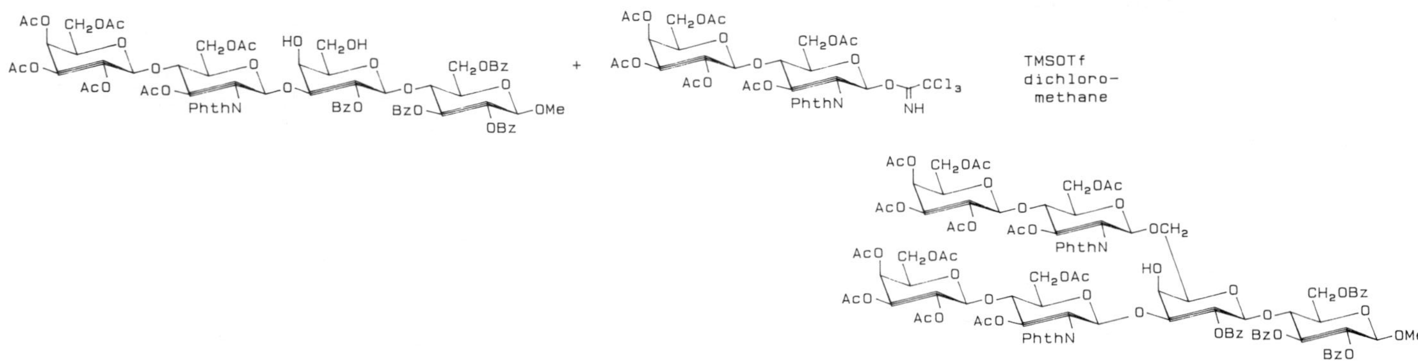

80%; m.p. 144–147 °C;
$[\alpha]_D$ +29.5 (chloroform); Ref.: 39

β-D-GlcpN-(1→2)-α-D-Manp-(1→3)
β-D-GlcpN-(1→2)-α-D-Manp-(1→6) 〉β-D-Manp-(1→4)-D-GlcN

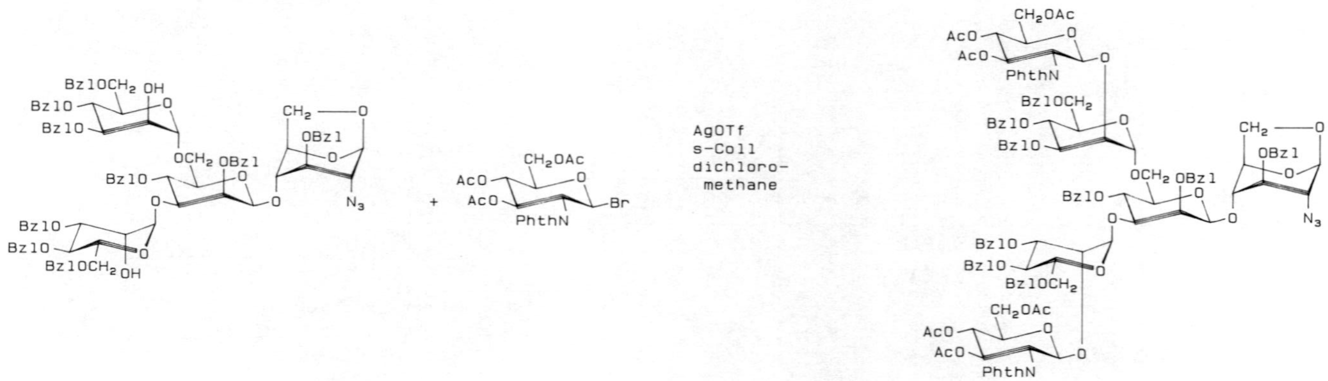

71%; syrup; $[\alpha]_D$ -5.0 (chloroform); Ref.: 157

β-D-Galp-(1→4)-β-D-GlcpNAc-(1→3)-β-D-Galp-(1→4)-β-D-GlcpNAc-(1→3)-β-D-Galp-(1→4)-D-GlcNAc

β-p-(1→OMe); amorphous powder; $[\alpha]_D$ -6 (water); Ref.: 80

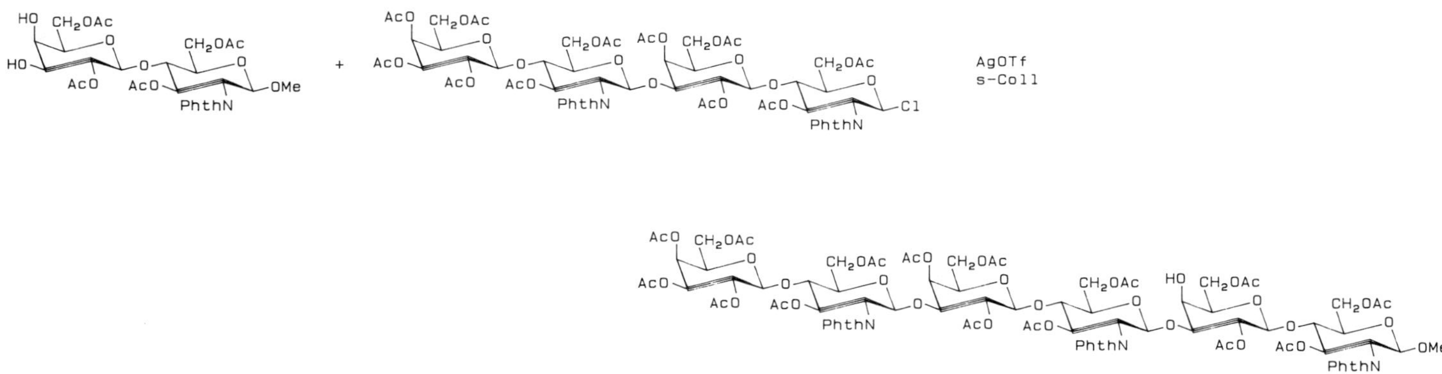

40%; m.p. 182–184 °C; $[\alpha]_D$ +16 (chloroform); Ref.: 80

α-D-Manp-(1→2)-α-D-Manp-(1→2)-α-D-Manp-(1→2)-α-D-Manp-(1→2)-α-D-Manp-(1→2)-D-Man

m.p. 177–182 °C; $[\alpha]_D$ +32.2 (water); C-13; Ref.: 97

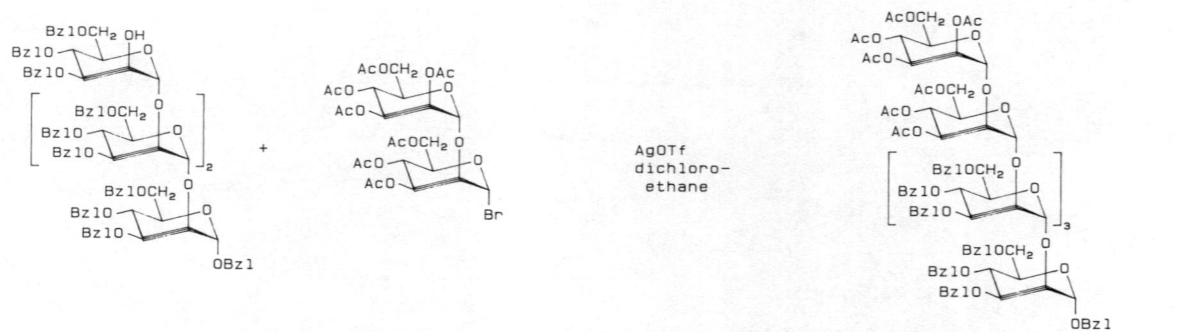

73%; $[\alpha]_D$ +21.1 (chloroform); Ref.: 97

α-D-Glcp-(1→2)-α-D-Glcp-(1→3)-α-D-Glcp-(1→3)-α-D-Manp-(1→2)-α-D-Manp-(1→2)-D-Man

$[\alpha]_D$ +104.8 (water); Ref.: 91

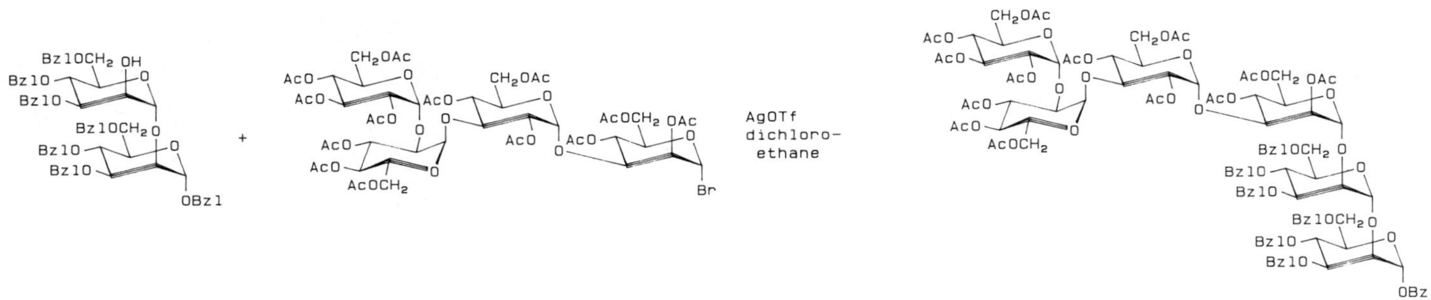

56%; $[\alpha]_D$ +75.0 (chloroform); Ref.: 91

β-D-Galp-(1→4)-β-D-GlcpNAc-(1→3)
β-D-Galp-(1→4)-β-D-GlcpNAc-(1→6) 〉α-D-Manp-(1→2)-D-Man

$[\alpha]_D$ −7 (water); C-13; Ref.: 175

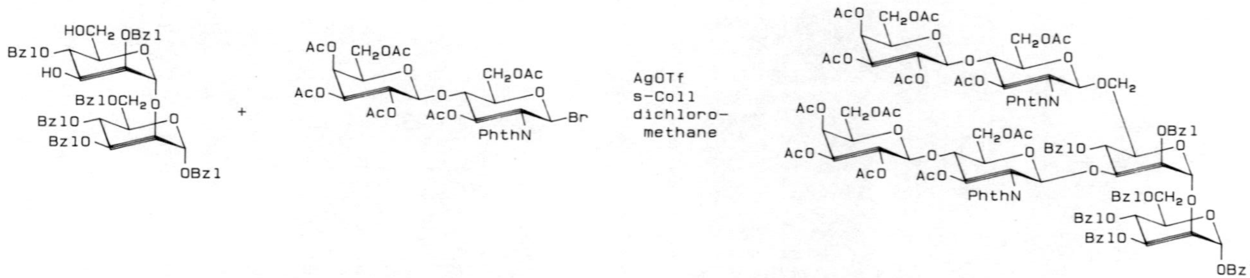

39%; syrup; $[\alpha]_D$ +0.5 (chloroform); C-13; Ref.: 175

α-D-Manp-(1→3)-α-D-Manp-(1→3)-α-D-Manp-(1→3)-α-D-Manp-(1→3)-α-D-Manp-(1→3)-D-Man

amorphous solid; $[\alpha]_D$ +26.8 (water); C-13; Ref.: 99

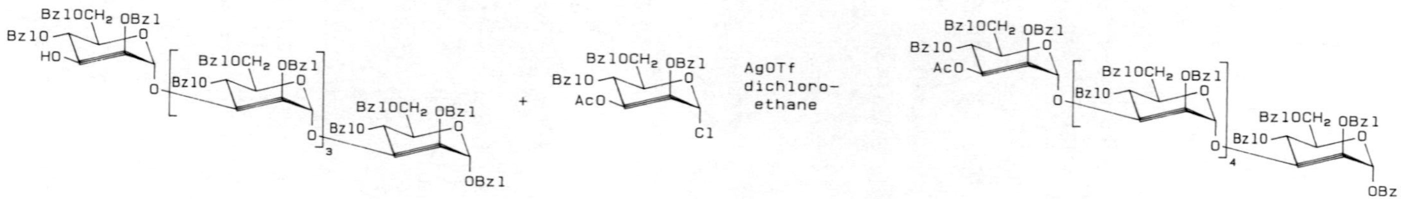

72%; $[\alpha]_D$ +11.0 (chloroform); Ref.: 99

α-D-Manp-(1→2)-α-D-Manp-(1→3)
α-D-Manp-(1→2)-α-D-Manp-(1→6) ⟩ α-D-Manp-(1→6)-D-Man

α-p-(1→OPr); $[\alpha]_D$ +69 (water); Ref.: 100

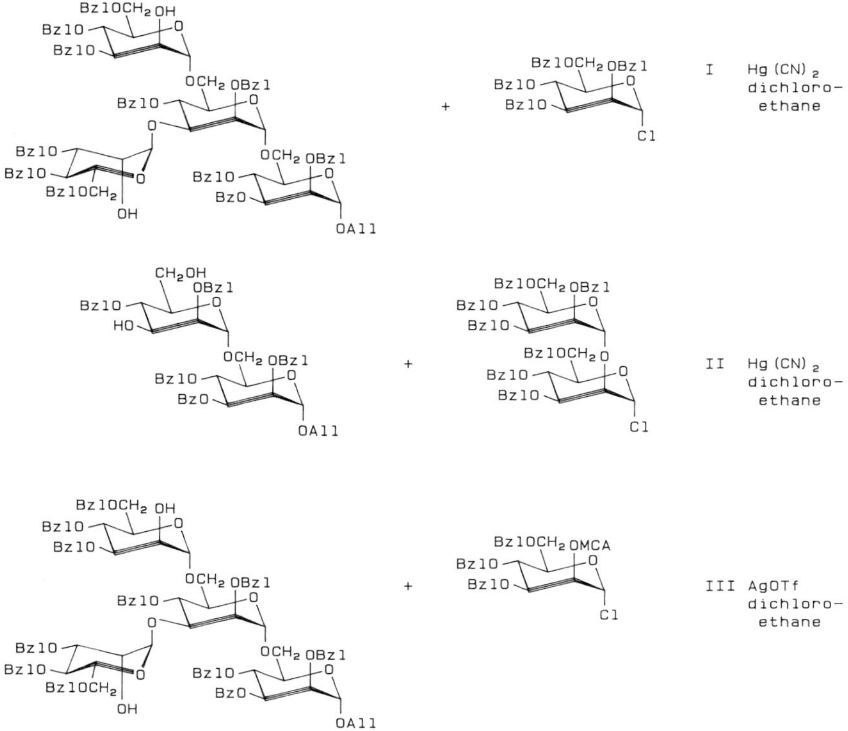

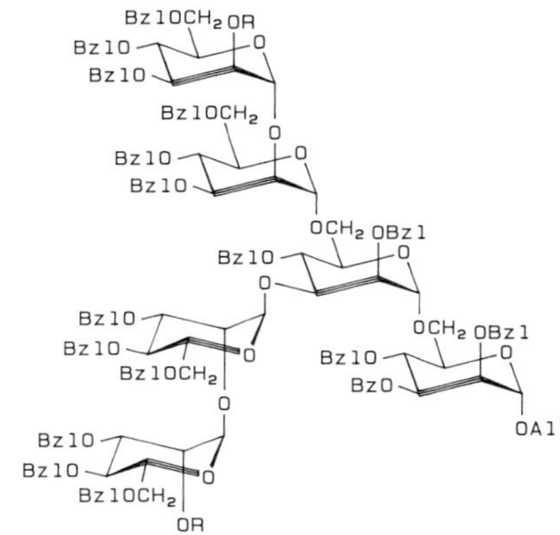

I R=Bzl; 29.3%; Ref.: 100

II R=Bzl; 8%;
$[\alpha]_D$ +20 (chloroform); C-13; Ref.: 100

III R=MCA; 54.7%;
$[\alpha]_D$ +16 (chloroform); C-13; Ref.: 100

α-D-Manp-(1→2)-α-D-Manp-(1→3)
α-D-Manp-(1→2)
α-D-Manp-(1→6)
α-D-Manp-(1→6)
D-Man

α-p-(1→OMe); amorphous; $[\alpha]_D$ +79.4 (water); C-13; Ref.: 171

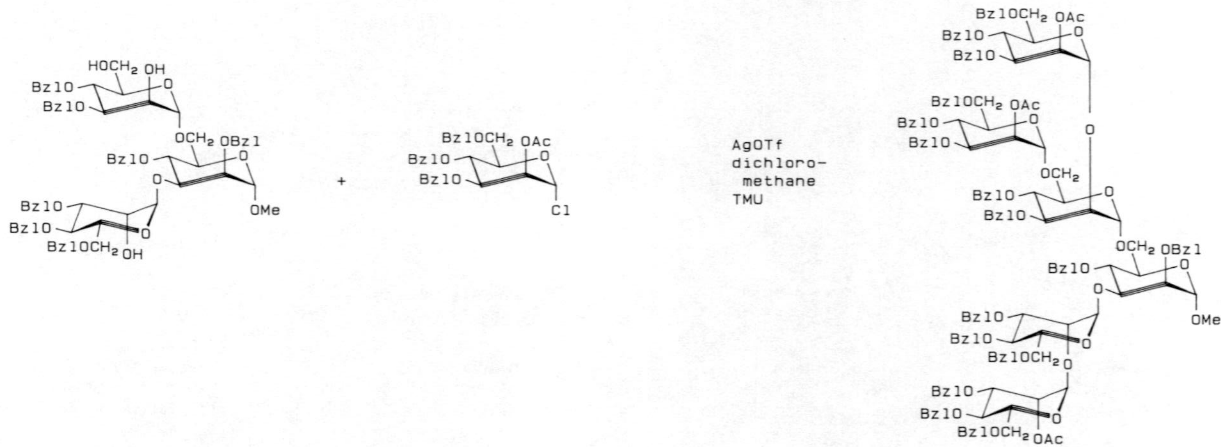

85.9%; $[\alpha]_D$ +41.6 (chloroform); Ref.: 171

α-D-Manp-(1→3)
α-D-Manp-(1→6) > α-D-Manp-(1→6) > D-Man
α-D-Manp-(1→2)-α-D-Manp-(1→3)

α-p-(1→OMe); $[\alpha]_D$ +92.0 (water); Ref.: 173

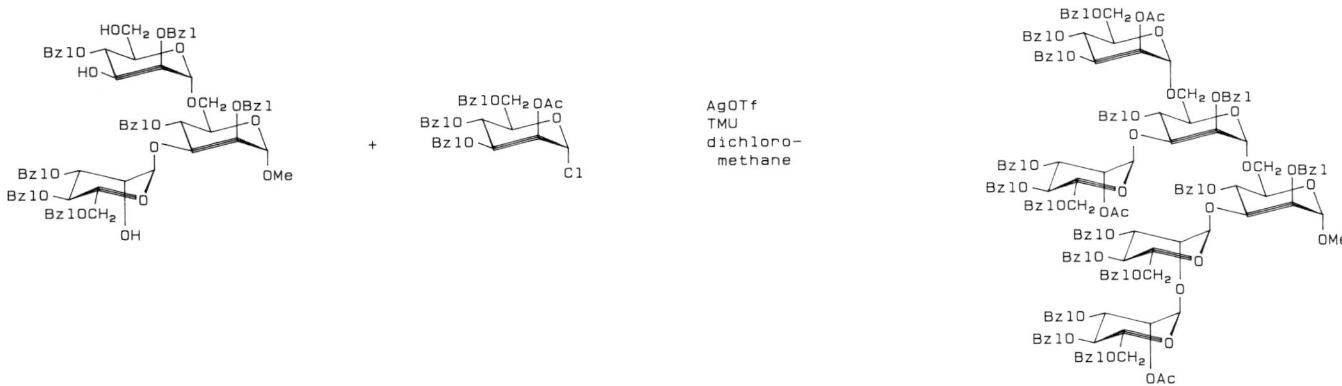

$[\alpha]_D$ +45.8; Ref.: 173

α-D-Manp-(1→3)
α-D-Manp-(1→6) 〉α-D-Manp-(1→3)
α-D-Manp-(1→2)-α-D-Manp-(1→6) 〉D-Man

α-p-(1→OMe); amorphous; $[\alpha]_D$ +102.1 (water); C-13; Ref.: 172

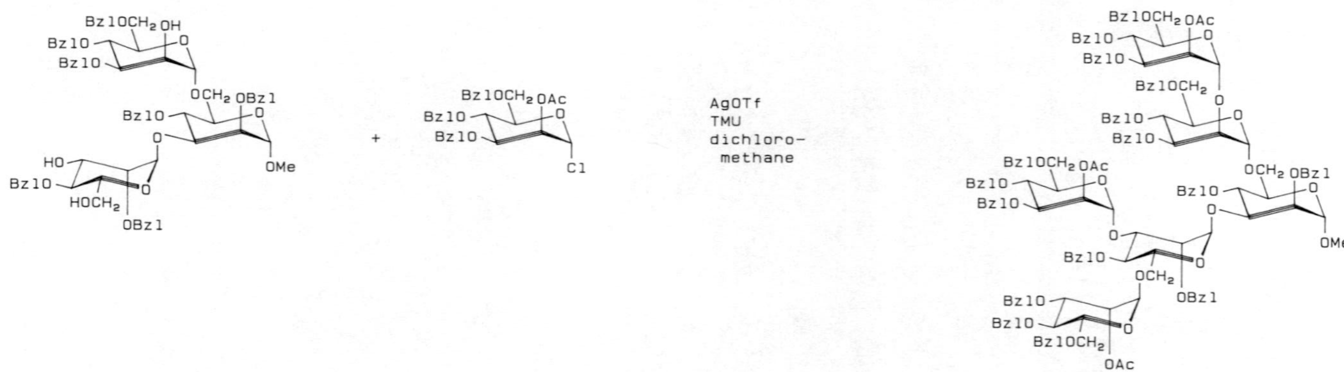

60 %; $[\alpha]_D$ +31 (chloroform); C-13; Ref.: 172

β-D-GlcpNAc-(1→2)-α-D-Manp-(1→3) ╲
　　　　　　　　　　　　　　　　　　D-Man
β-D-GlcpNAc-(1→2) ╲
　　　　　　　　　　α-D-Manp-(1→6) ╱
β-D-GlcpNAc-(1→4) ╱

α-p-(1→OMe); $[\alpha]_D$ +14.7 (water); C-13; Ref.: 176, 177

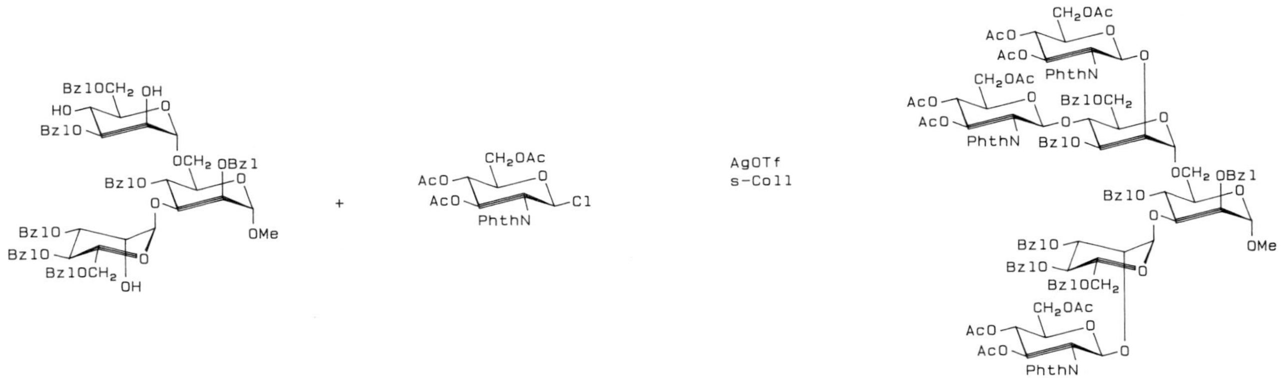

34.7%; $[\alpha]_D$ +22.4 (chloroform); Ref.: 176, 177

β-D-GlcpNAc-(1→2)-α-D-Manp-(1→3)
β-D-GlcpNAc-(1→4) → D-Man
β-D-GlcpNAc-(1→2)-α-D-Manp-(1→6)

$[\alpha]_D$ +26 (water); C-13; Ref.: 105

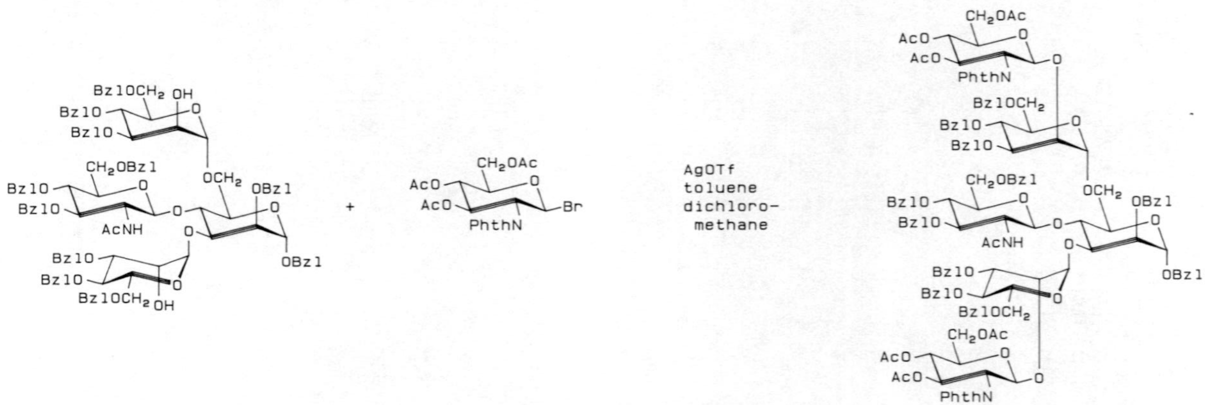

52%; syrup; $[\alpha]_D$ +21 (chloroform); C-13; Ref.: 105

β-D-Glcp-(1→4)-α-L-Rhap-(1→3)-β-D-Galp-(1→6)-β-D-Glcp-(1→4)-α-L-Rhap-(1→3)-D-Gal

$[\alpha]_D$ -18 (water); C-13; Ref.: 50

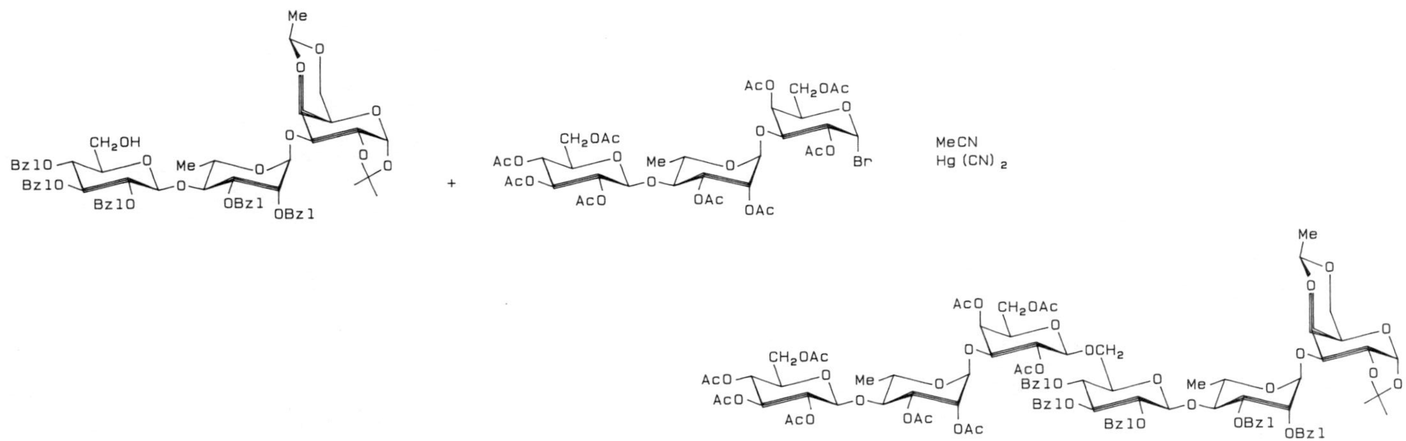

60%; syrup; $[\alpha]_D$ -7 (chloroform); Ref.: 50

β-D-Manp-(1→4)-α-L-Rhap-(1→3)-β-D-Galp-(1→6)-β-D-Manp-(1→4)-α-L-Rhap-(1→3)-D-Gal

$[\alpha]_D$ -32.5 (water); C-13; Ref.: 178, 179, 180

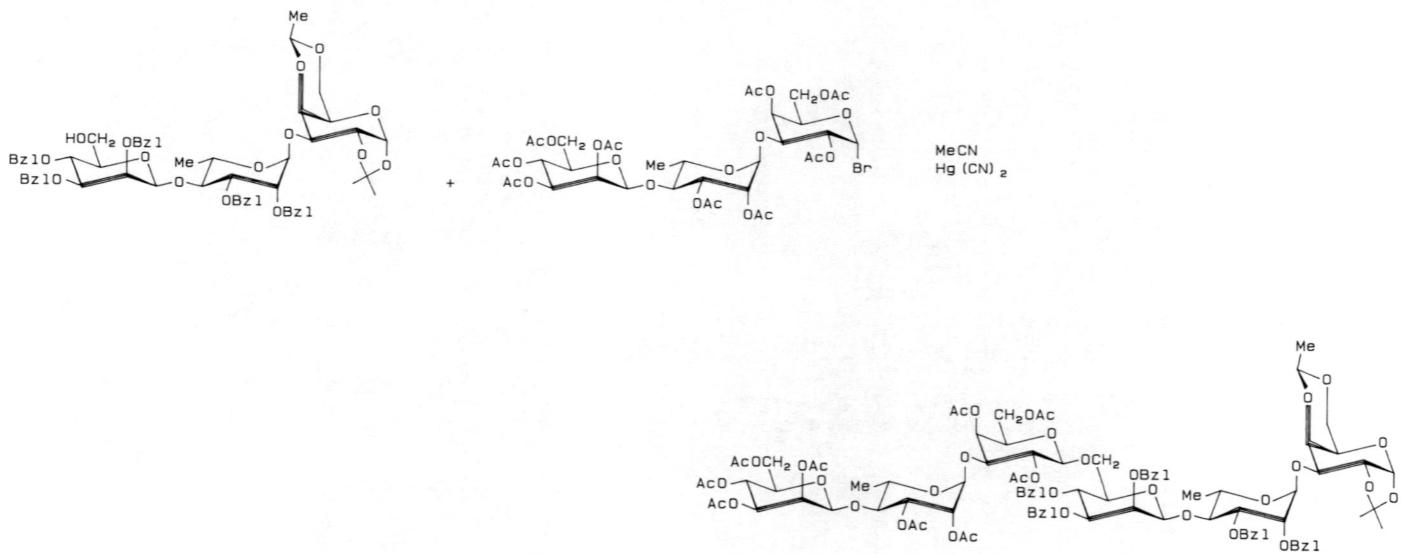

84%; solid; $[\alpha]_D$ -13.5 (chloroform); Ref.: 178, 179, 180

β-D-Galp-(1→6)-β-D-Galp-(1→6)-β-D-Galp-(1→6)-β-D-Galp-(1→6)-β-D-Galp-(1→6)-D-Gal

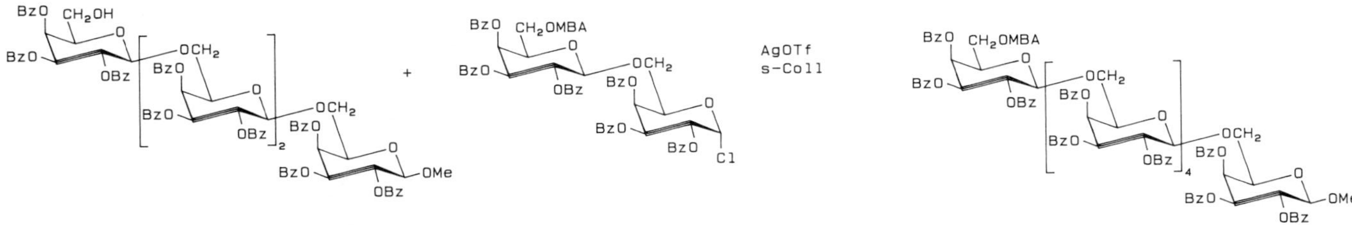

66 %; m.p. 303.5-304.5 °C;
$[\alpha]_D$ +77 (chloroform); Ref.: 137

HEPTASACCHARIDES

Having monosaccharide units at the reducing end in the following sequence:

D-Glc
D-GlcN
D-Man
D-Gal

β-D-Glcp-(1→3)-[β-D-Glcp-(1→6)]-β-D-Glcp-(1→6)-β-D-Glcp-(1→6)-[β-D-Glcp-(1→3)]-β-D-Glcp-(1→6)-D-Glc

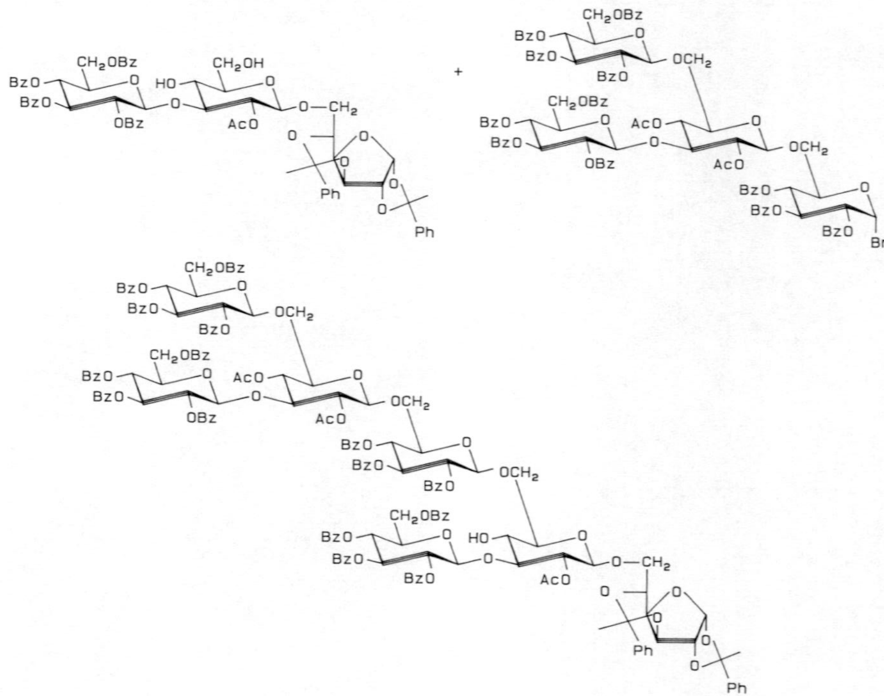

AgOTf

59 %; Ref.: 20, 21

α-D-Glcp-(1→3)-α-D-Glcp-(1→6)-α-D-Glcp-(1→6)
α-D-Glcp-(1→3)-α-D-Glcp-(1→6)-α-D-Glcp-(1→3) D-Glc

$[\alpha]_D$ +196 (water), C-13, Ref: 181

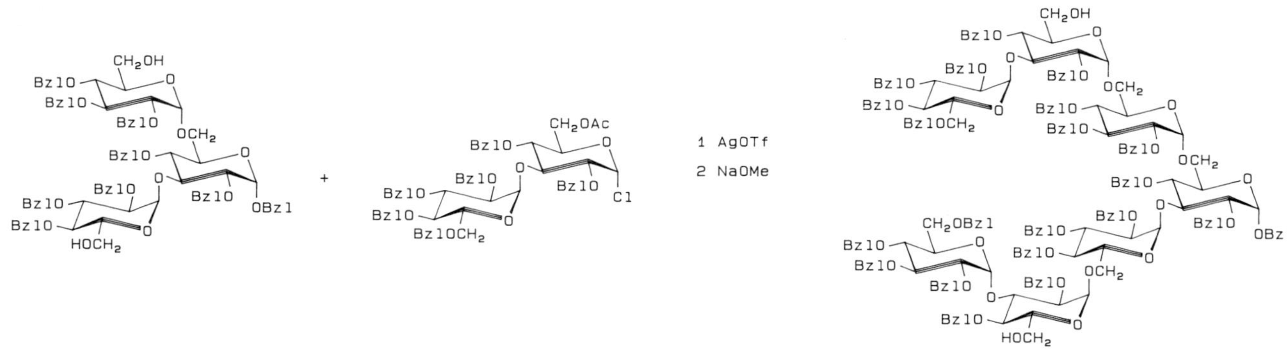

36%; $[\alpha]_D$ +93; Ref.: 181

α-Neu5Ac-(2-6)-β-D-Galp-(1-4)-β-D-GlcpNAc-(1-2)-α-D-Manp-(1-6)-β-D-Manp-(1-4)-β-D-GlcpNAc-(1-4)-D-GlcNAc

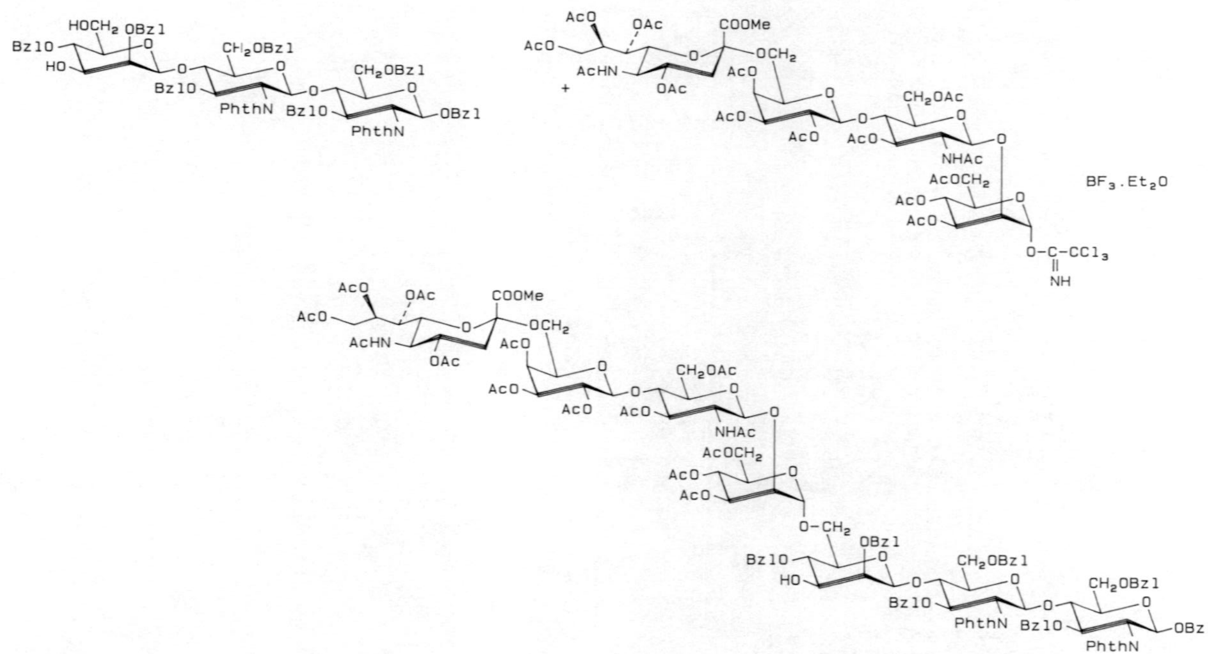

48 %
$[\alpha]_D$ -12.2 (chloroform); Ref.: 95

β-D-Galp-(1→4)-β-D-GlcpNAc-(1→2)-α-D-Manp-(1→6)
β-D-Galp-(1→4)-β-D-GlcpNAc-(1→2)-α-D-Manp-(1→3) 〉D-Man

amorphous powder; $[\alpha]_D$ +14 (water); C-13; Ref.: 161, 162

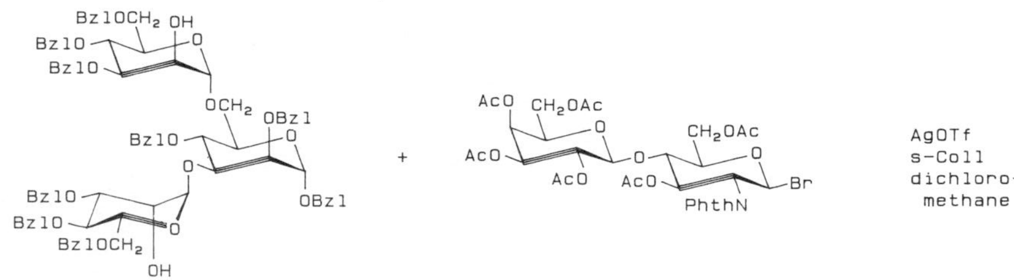

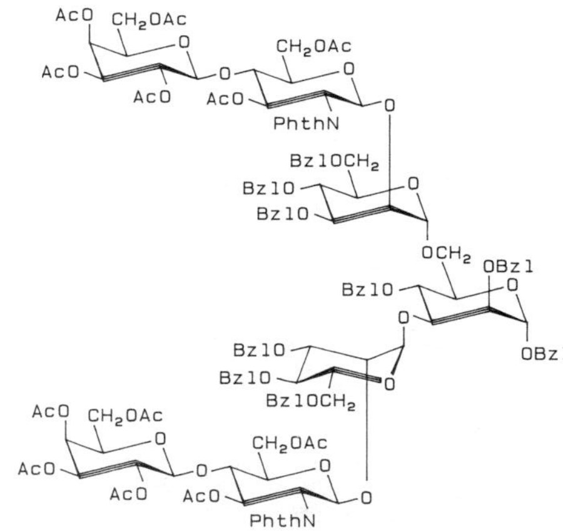

56%; syrup; $[\alpha]_D$ +10 (chloroform); C-13; Ref.: 161, 162

α-L-Fucp-(1→3)-β-D-GlcpNAc-(1→2)-α-D-Manp-(1→6)
α-L-Fucp-(1→3)-β-D-GlcpNAc-(1→2)-α-D-Manp-(1→3) 〉D-Man

amorphous powder, $[\alpha]_D$ -44 (water), C-13, Ref.: 182

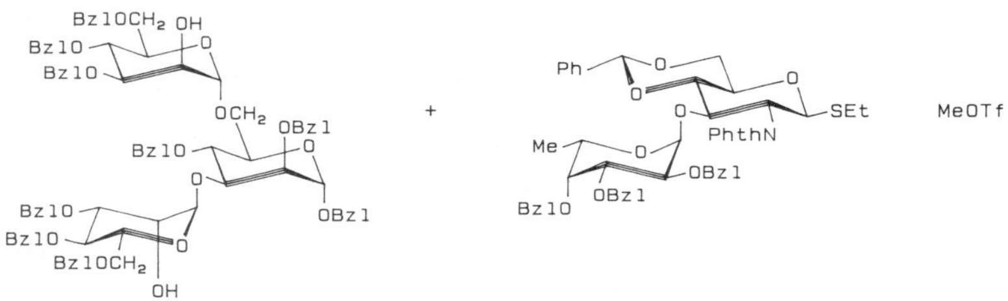

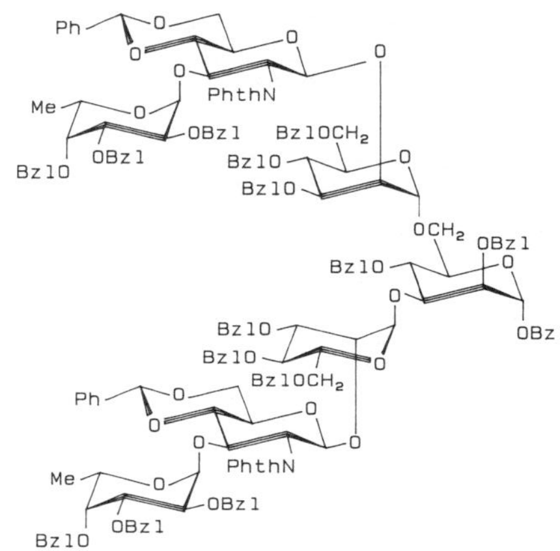

63%; syrup;
$[\alpha]_D$ -7 (chloroform); C-13; Ref.: 182

β-D-Galp-(1→4)-β-D-GlcpNAc-(1→2)
β-D-Galp-(1→4)-β-D-GlcpNAc-(1→4) 〉α-D-Manp-(1→3)
α-D-Manp-(1→6) 〉D-Man

β-p-(1→OECO); amorphous powder; $[\alpha]_D$ +4 (water), Ref.: 168

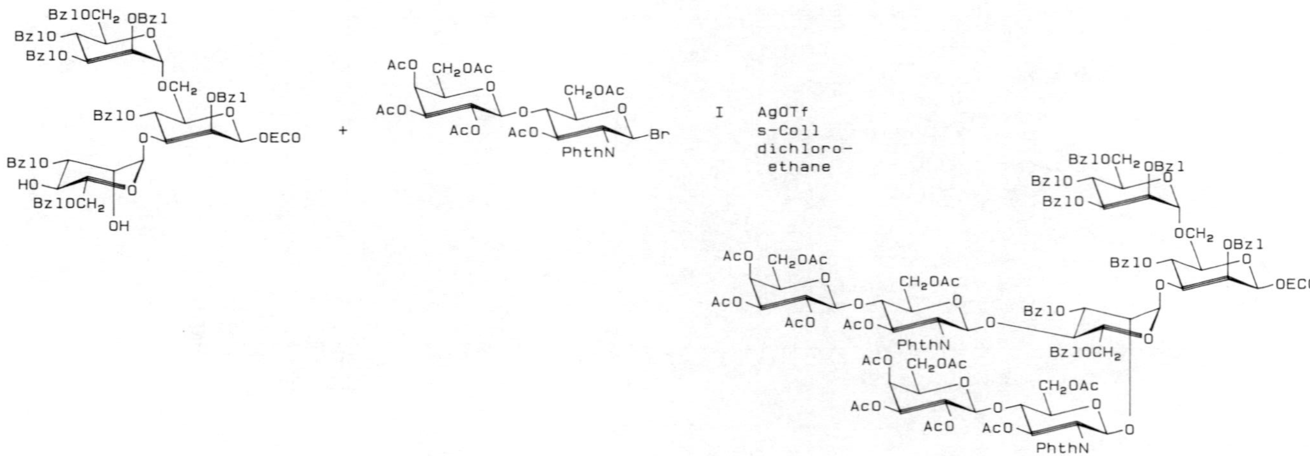

I 37%; $[\alpha]_D$ +2.5 (chloroform); Ref.: 168

I 38%; $[\alpha]_D$ +2.5 (chloroform); Ref.: 169

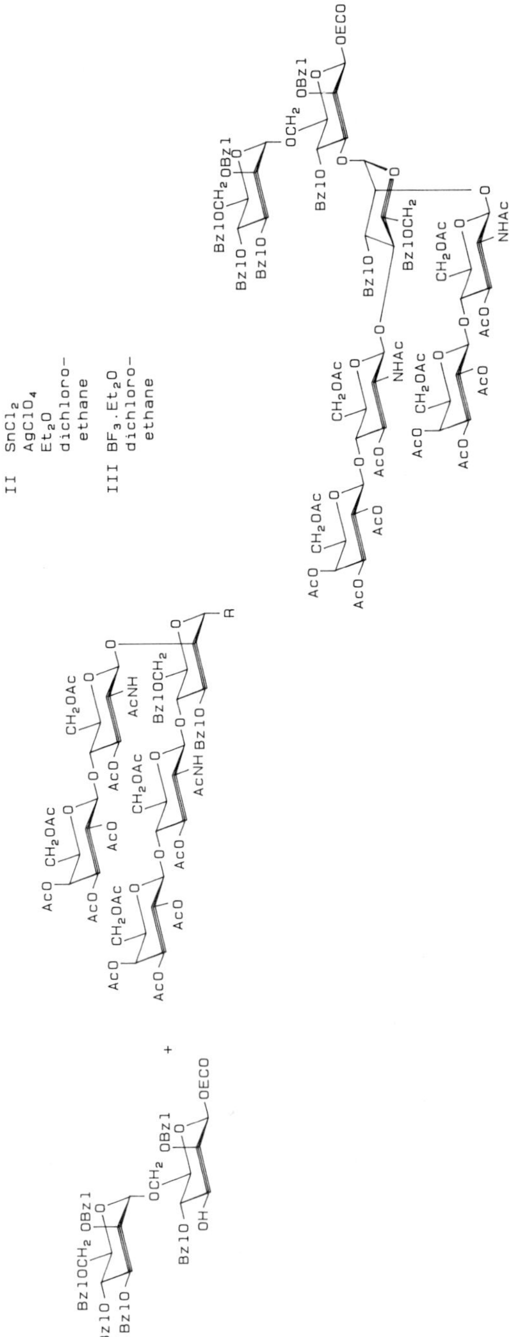

II R=F, 13%; Ref.: 164

III R=O-CNH-CCl$_3$; 4.6%; $[\alpha]_D$ -6.6 (chloroform); Ref.: 164

β-D-Galp-(1→4)-β-D-GlcpNAc-(1→2) \
                                 〉β-D-Manp-(1→3) \
β-D-Galp-(1→4)-β-D-GlcpNAc-(1→4) /   〉D-Man
                            α-D-Manp-(1→6) /

β-p-(1-OMCO); Ref.: 164

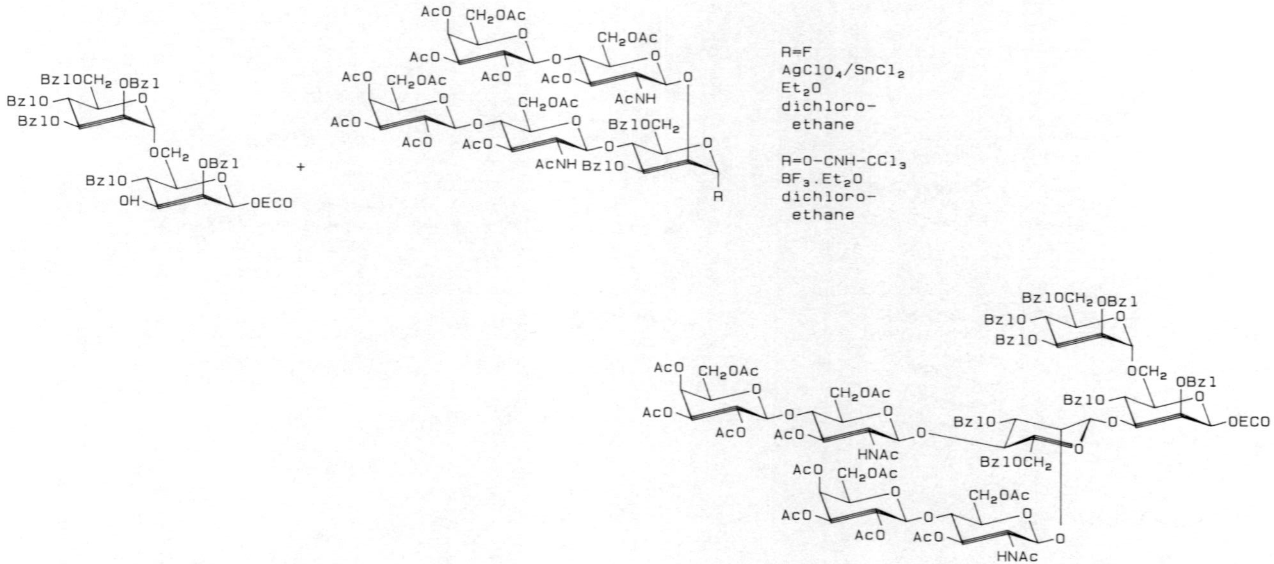

R=F; 13%; Ref.: 164

R=O-CNH-CCl_3; 2%; $[\alpha]_D$ -10 (chloroform); Ref.: 164

β-D-Galp-(1→3)-β-D-Galp-(1→3)-β-D-Galp-(1→3)-β-D-Galp-(1→3)-β-D-Galp-(1→3)-β-D-Galp-(1→3)-D-Gal

β-p-(1→OMe); $[\alpha]_D$ +36.2 (water); amorphous hygroscopic solid; C-13; Ref.: 132

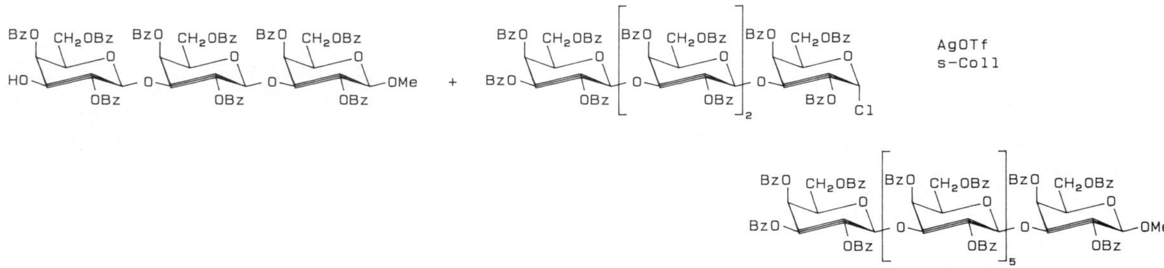

62%; amorphous; $[\alpha]_D$ +33.8 (chloroform); Ref.: 132

OCTASACCHARIDES

Having monosaccharide units at the reducing end in the following sequence:

D-Glc
D-GlcN (D-GlcNAc)
D-Man
L-Rha

α-D-Glcp-(1→6)-α-D-Glcp-(1→6)-α-D-Glcp-(1→6)-α-D-Glcp-(1→6)-α-D-Glcp-(1→6)-α-D-Glcp-(1→6)-α-D-Glcp-(1→6)-D-Glc

glass; $[\alpha]_D$ +165 (water); Ref.: 8, 9

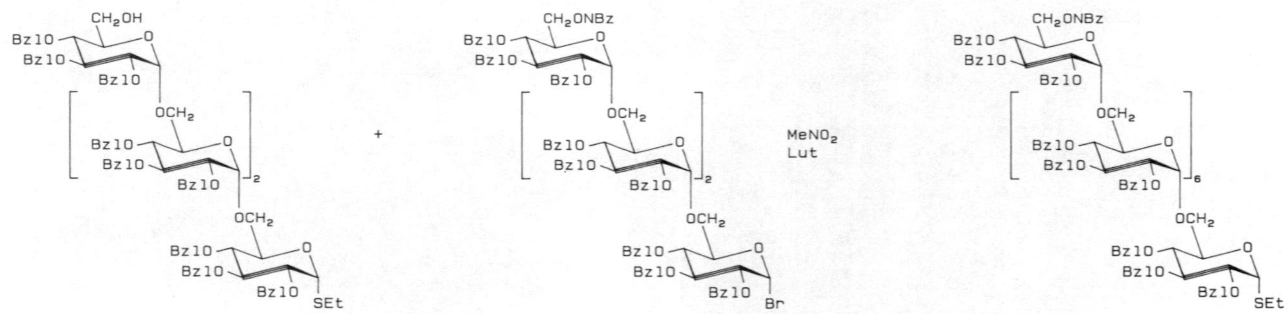

11%; $[\alpha]_D$ +110 (chloroform); Ref.: 8, 9

β-D-Glcp-(1→3)-β-D-Glcp-(1→3)
β-D-Glcp-(1→3) \ β-D-Glcp-(1→6)-β-D-Glcp-(1→6) \ β-D-Glcp-(1→6)-D-Glc
β-D-Glcp-(1→6) /

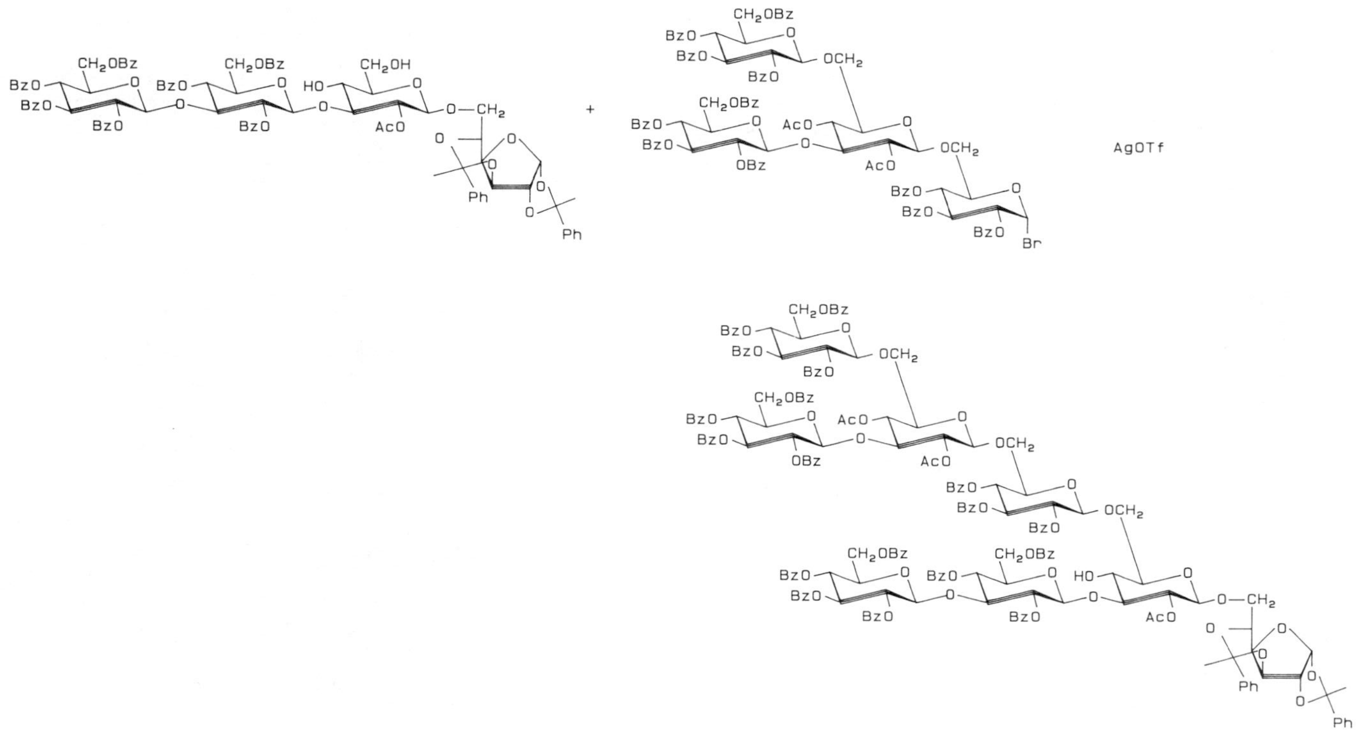

48%; Ref.: 20, 21

β-D-Galp-(1→4)-β-D-GlcpNAc-(1→2)-α-D-Manp-(1→6)
β-D-Galp-(1→4)-β-D-GlcpNAc-(1→2)-α-D-Manp-(1→3) 〉β-D-Manp-(1→4)-D-GlcNAc

$[\alpha]_D$ +1.8 (water); Ref.: 157, 158

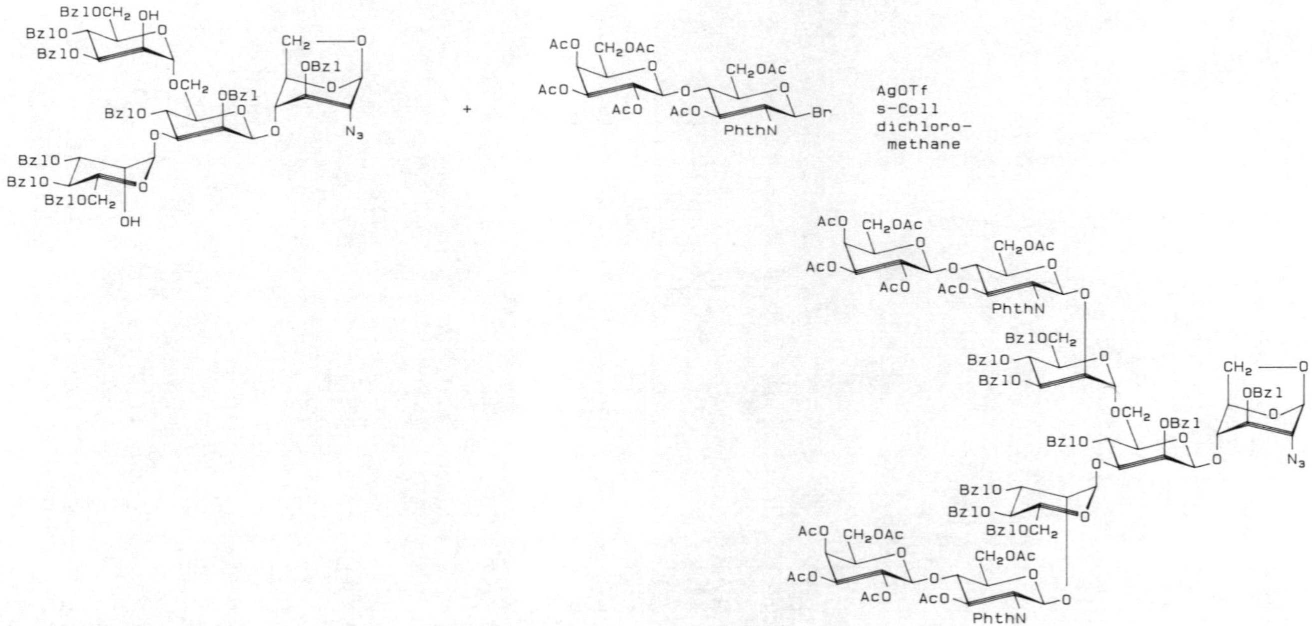

70%; $[\alpha]_D$ −8 (chloroform); Ref.: 157, 158

β-D-Galp-(1→4)-β-D-GlcpNAc-(1→2)-α-D-Manp-(1→6)
β-D-GlcpNAc-(1→4) → D-Man
β-D-Galp-(1→4)-β-D-GlcpNAc-(1→2)-α-D-Manp-(1→3)

$[\alpha]_D$ +18 (water); C-13; Ref.: 105

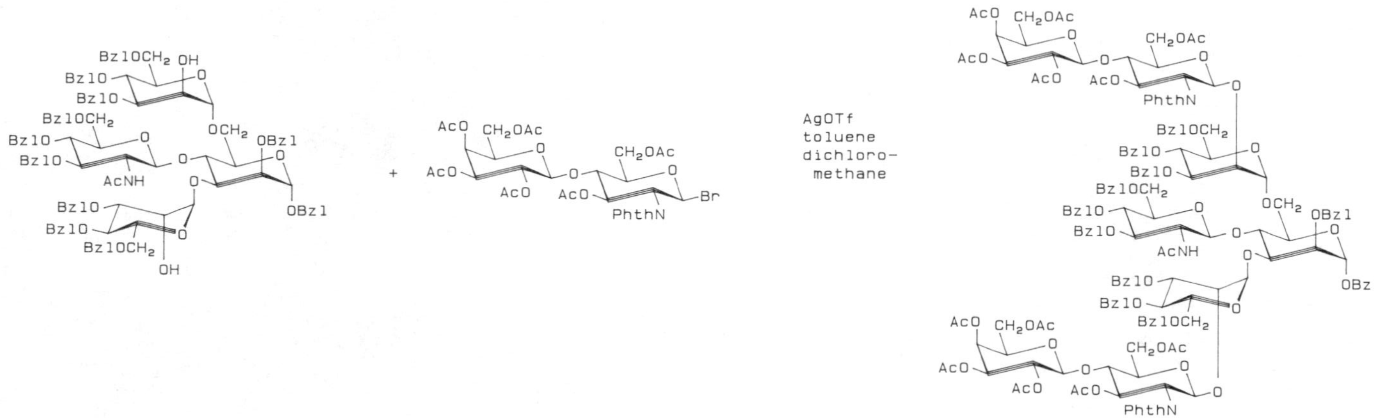

57%; $[\alpha]_D$ +12 (chloroform); C-13; Ref.: 105

β-D-GlcpNAc-(1→2)-α-L-Rhap-(1→2)-α-L-Rhap-(1→3)-α-L-Rhap-(1→3)-β-D-GlcpNAc-(1→2)-α-L-Rhap-(1→2)-α-L-Rhap-(1→3)-L-Rha

α-p-(1→OMe); $[\alpha]_D$ −63 (water); C-13; Ref.: 116

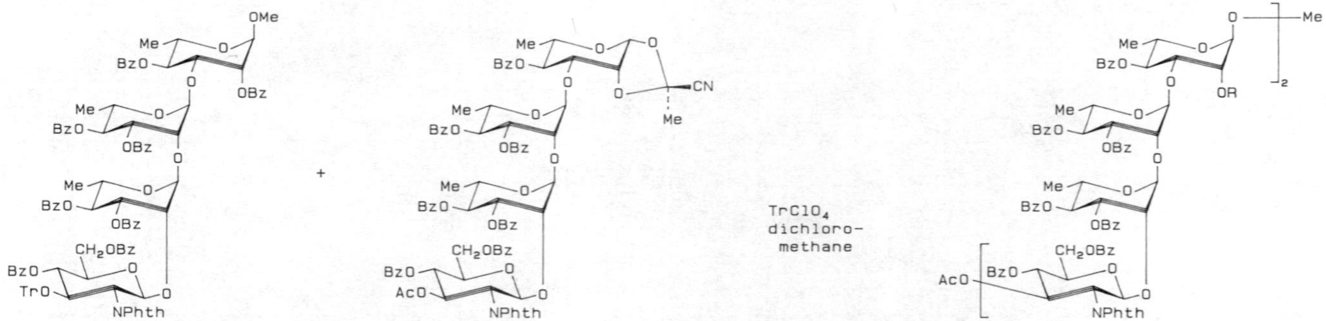

R=Bz, Ac

94%; syrup; $[\alpha]_D$ +120.5 (chloroform); Ref.: 116

NONASACCHARIDES

Having monosaccharide units at the reducing end in the following sequence:

D-GlcN (D-GlcNAc)
D-Man
D-Gal

β-D-Galp-(1→4)-β-D-GlcpNAc-(1→2)-α-D-Manp-(1→3) \
β-D-Galp-(1→4)-β-D-GlcpNAc-(1→2)-α-D-Manp-(1→6) / β-D-Manp-(1→4)-β-D-GlcpNAc-(1→4)-D-GlcNAc

$[\alpha]_D$ +1.8 (water); Ref.: 183

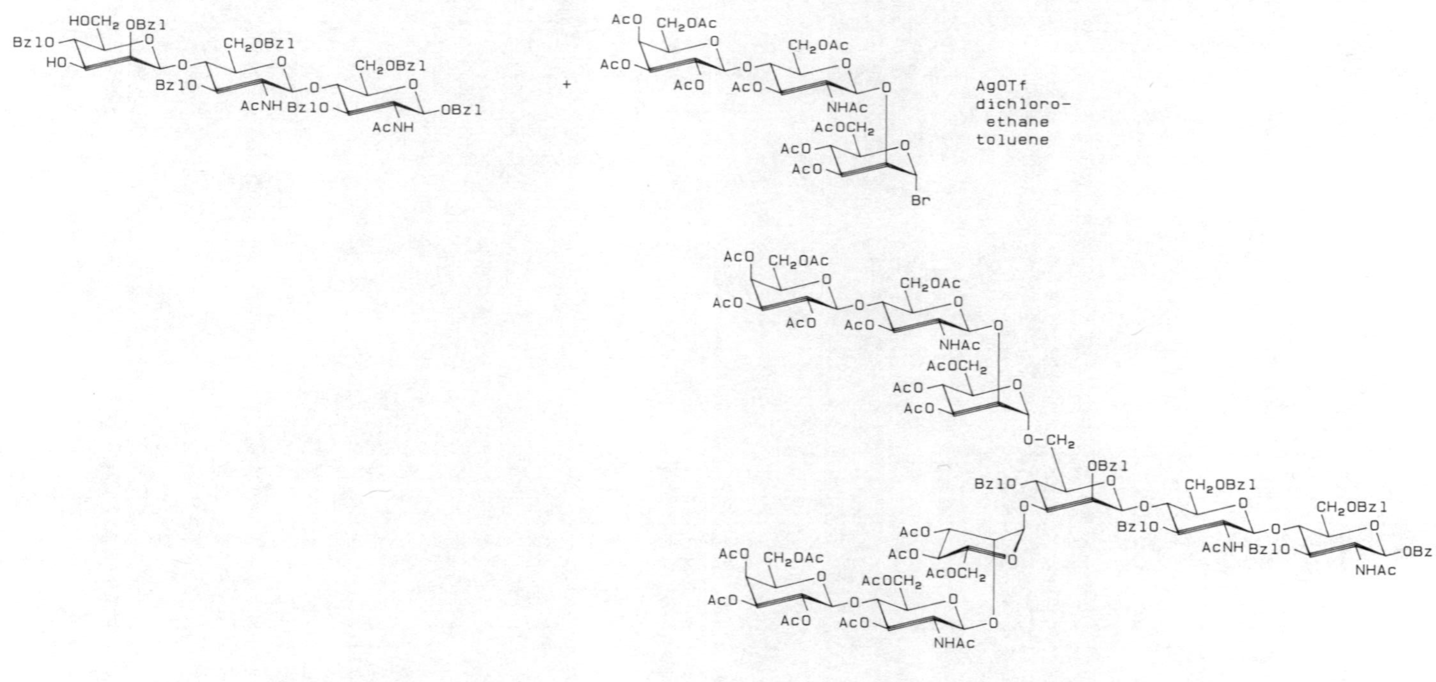

β-D-Galp-(1→4) ⟩ β-D-GlcpNAc-(1→2)-α-D-Manp-(1→6) ⟩ D-Man
α-L-Fucp-(1→3)
β-D-Galp-(1→4) ⟩ β-D-GlcpNAc-(1→2)-α-D-Manp-(1→3)
α-L-Fucp-(1→3)

$[\alpha]_D$ -34 (water); Ref.: 96

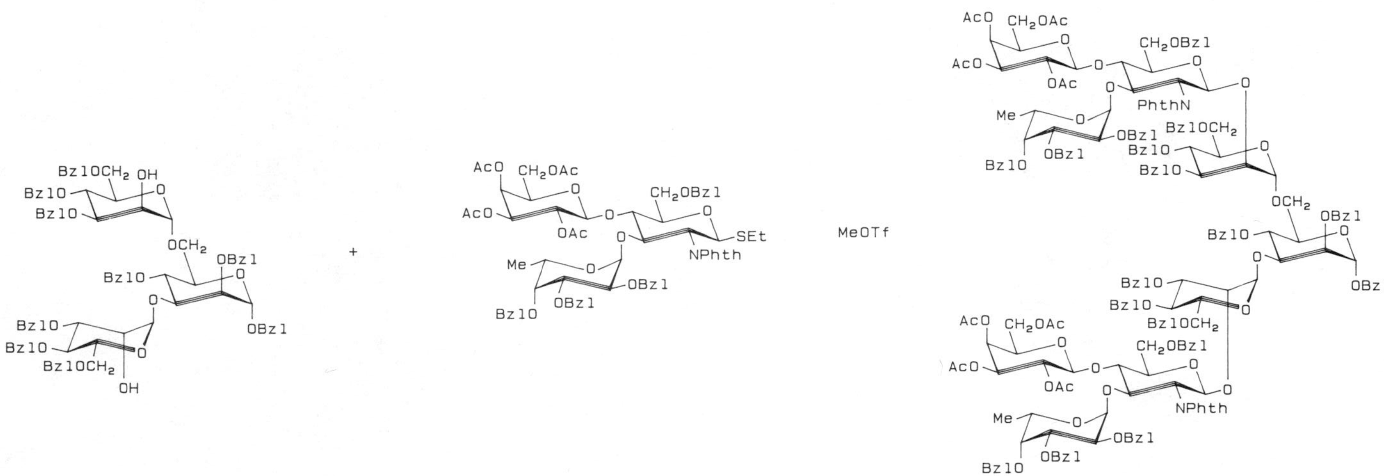

61%; $[\alpha]_D$ -5 (dichloromethane); Ref.: 96

β-D-Galp-(1→4)-β-D-GlcpNAc-(1→2)
β-D-Galp-(1→4)-β-D-GlcpNAc-(1→4) ⟩ α-D-Manp-(1→3)
β-D-Galp-(1→4)-β-D-GlcpNAc-(1→2)-α-D-Manp-(1→6) ⟩ D-Man

amorphous powder; $[\alpha]_D$ +8 (water); C-13; Ref.: 184

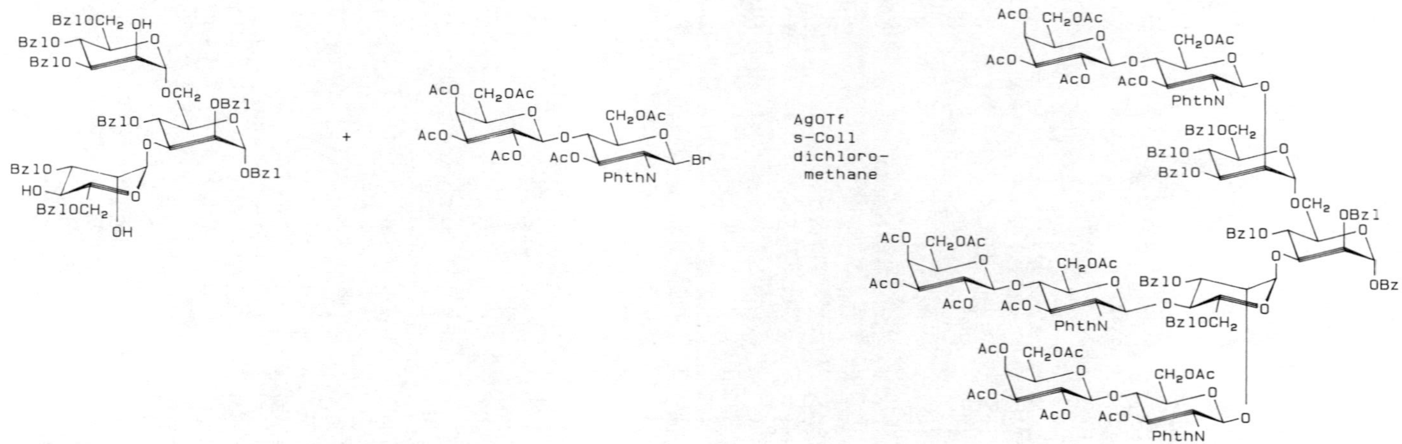

19%; amorphous powder;
$[\alpha]_D$ +11 (chloroform); C-13; Ref.: 184

β-D-Galp-(1→4)-β-D-GlcpNAc-(1→4)-α-D-Manp-(1→3) \
β-D-Galp-(1→4)-β-D-GlcpNAc-(1→2) \ D-Man
　　　　　　　　　　　　　　　　　　α-D-Manp-(1→6) /
β-D-Galp-(1→4)-β-D-GlcpNAc-(1→6) /

amorphous powder, $[\alpha]_{578}$ +5 (water), C-13, Ref.: 185

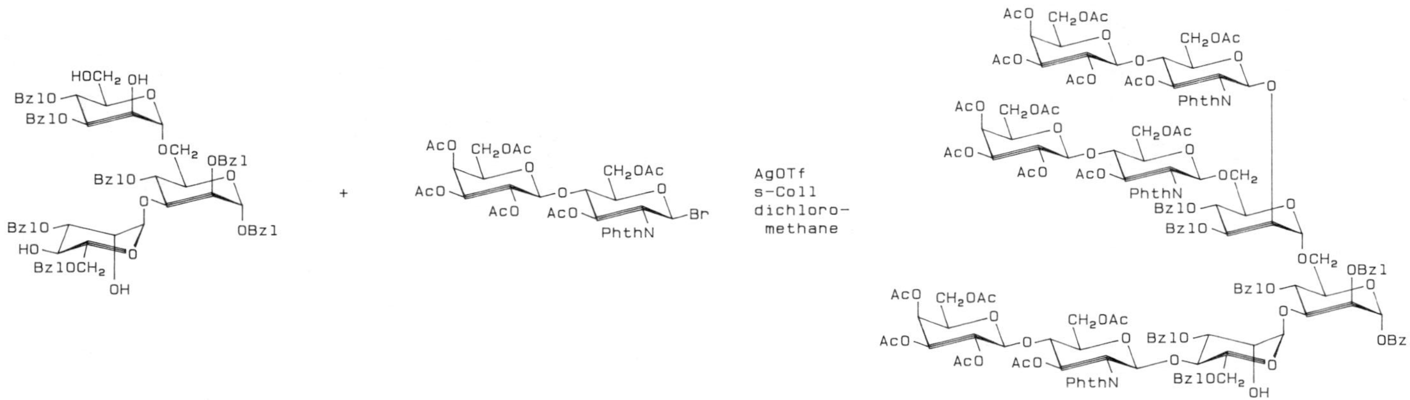

29%; syrup;
$[\alpha]_{578}$ +19 (chloroform); C-13; Ref.: 185

β-D-Manp-(1→4)-α-L-Rhap-(1→3)-β-D-Galp-(1→6)-β-D-Manp-(1→4)-α-L-Rhap-(1→3)-β-D-Galp-(1→6)-β-D-Manp-(1→4)-α-L-Rhap-(1→3)-D-Gal

[α]$_D$ −32 (water); C-13; Ref.: 178, 179, 180

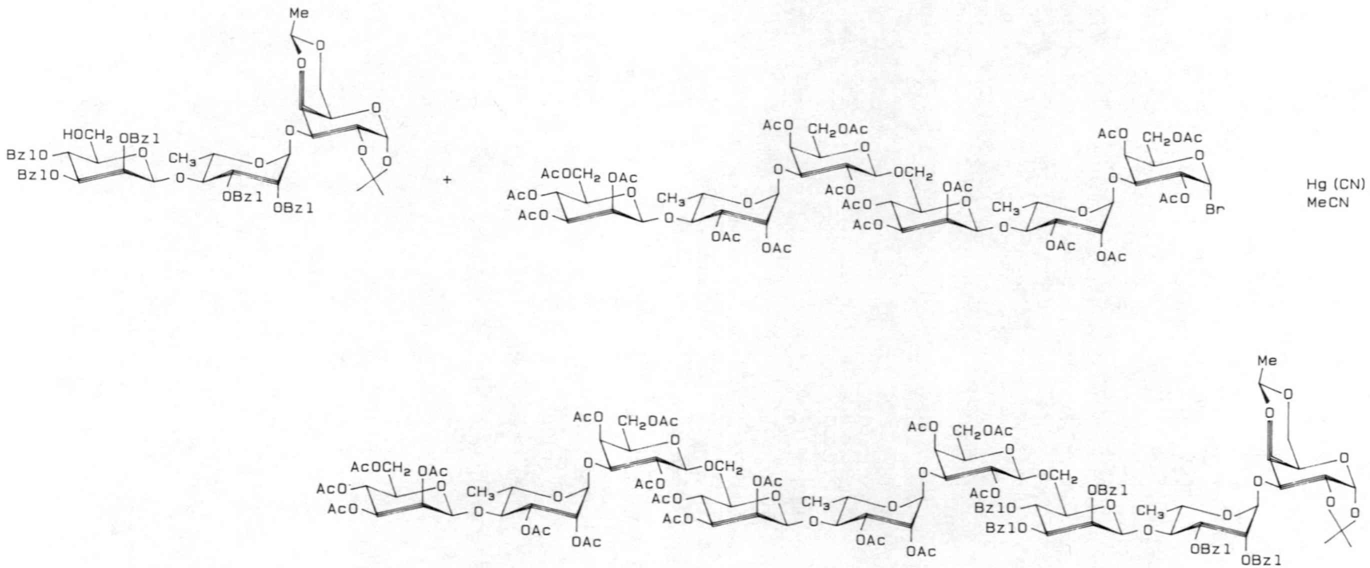

57%; solid; [α]$_D$ −12 (chloroform); Ref.: 178, 179, 180

UNDECASACCHARIDES

Having monosaccharide units at the reducing end in the following sequence:

D-GlcN (D-GlcNAc)
D-Man

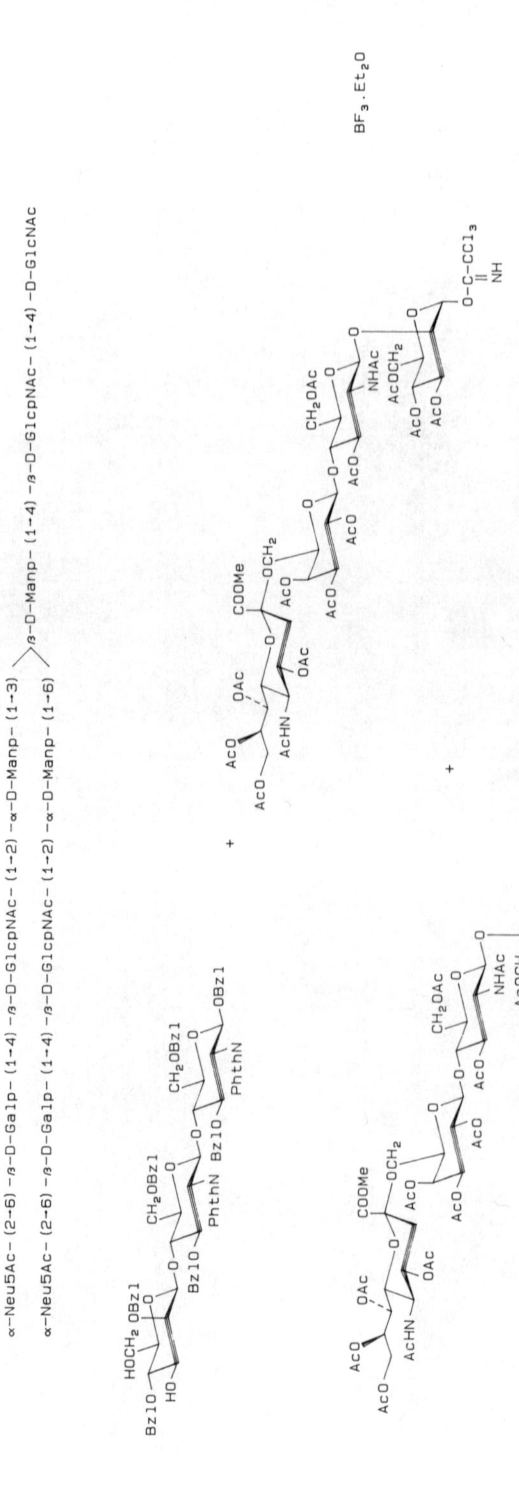
α-Neu5Ac-(2→6)-β-D-Galp-(1→4)-β-D-GlcpNAc-(1→2)-α-D-Manp-(1→3)
α-Neu5Ac-(2→6)-β-D-Galp-(1→4)-β-D-GlcpNAc-(1→2)-α-D-Manp-(1→6)
β-D-Manp-(1→4)-β-D-GlcpNAc-(1→4)-D-GlcNAc
BF3.Et2O

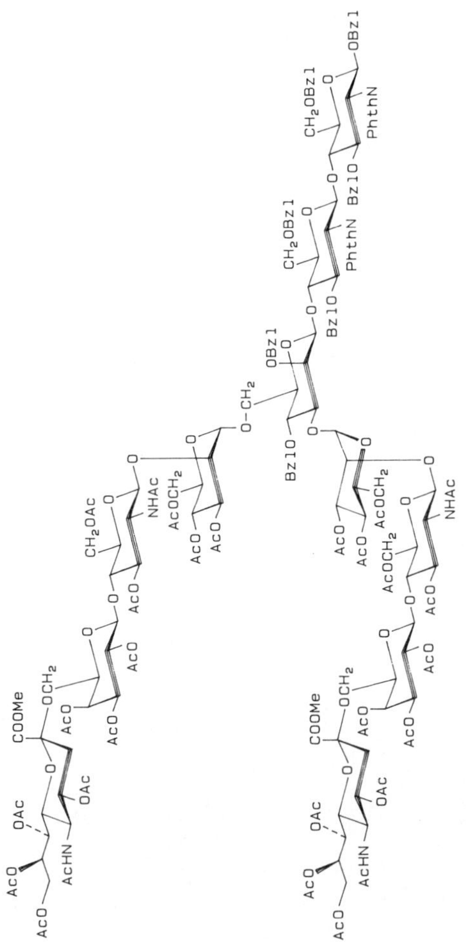

56%; $[\alpha]_D$ -7.8 (chloroform); Ref.: 95

β-D-Galp-(1→4)-β-D-GlcpNAc-(1→2)
β-D-Galp-(1→4)-β-D-GlcpNAc-(1→4) 〉α-D-Manp-(1→3)
〉D-Man
β-D-Galp-(1→4)-β-D-GlcpNAc-(1→2)
β-D-Galp-(1→4)-β-D-GlcpNAc-(1→6) 〉α-D-Manp-(1→6)

amorphous powder; $[\alpha]_{578}$ +3 (water); C-13; Ref.: 185

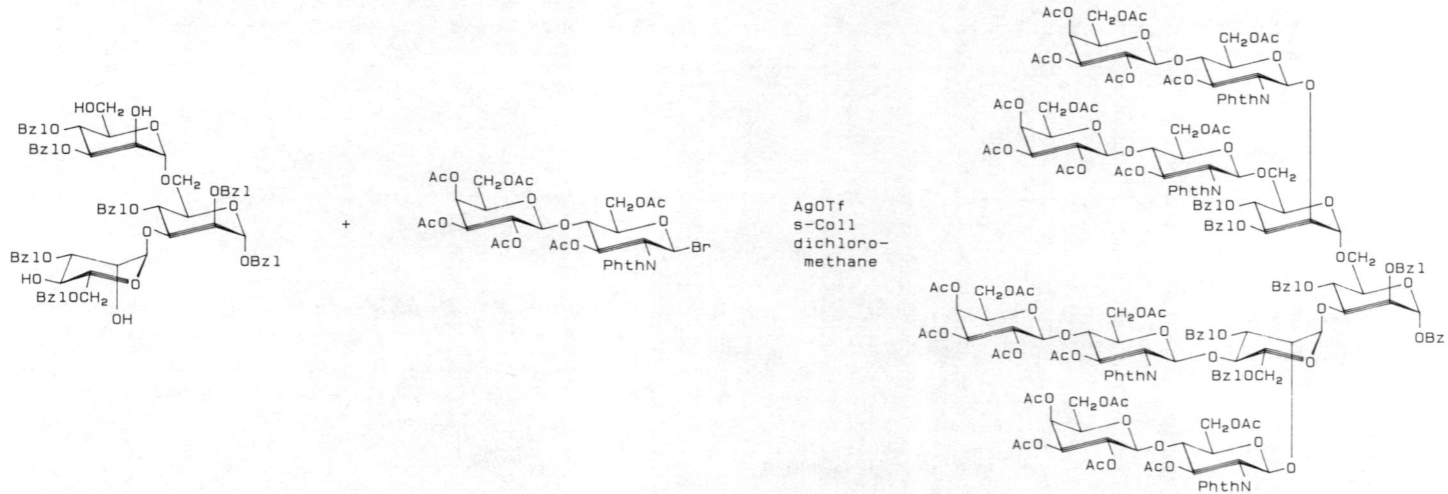

syrup; impure; C-13; Ref.: 185

REFERENCES

1. **Dupeyre, D., Excoffier, G., and Utille, J.-P.,** Stepwise synthesis of linear β-D-(1→3)-xylo-oligosaccharides. Preparation of a β,β,β-D-linked tetrasaccharide derivative, *Carbohydr. Res.*, 135, C1, 1984.
2. **Kováč, P. and Hirsch, J.,** Sequential synthesis and ^{13}C-N.M.R. spectra of methyl β-glycosides of (1→4)-β-D-xylo-oligosaccharides, *Carbohydr. Res.*, 100, 177, 1982.
3. **Hirsch, J., Kováč, P., and Petráková, E.,** An approach to the systematic synthesis of (1→4)-β-D-xylo-oligosaccharides, *Carbohydr. Res.*, 106, 203, 1982.
4. **Kováč, P. and Hirsch, J.,** Systematic, sequential synthesis of (1→4)-β-D-xylo-oligosaccharides and their methyl β-glycosides, *Carbohydr. Res.*, 90, C5, 1981.
5. **Kováč, P., Hirsch, J., and Kováčik, V.,** The stepwise synthesis of a methyl β-xylotetraoside related to branched xylans, *Carbohydr. Res.*, 75, 109, 1979.
6. **Ogawa, T. and Takahashi, Y.,** Total synthesis of α-cyclodextrin, *Carbohydr. Res.*, 138, C5, 1985.
7. **Backinowsky, L. V., Tsvetkov, Yu. E., and Kochetkov, N. K.,** Tritylated maltose 1,2-thioorthoesters in polycondensation reaction, *Bioorg. Khim.*, 7, 750, 1981.
8. **Koto, S., Uchida, T., and Zen, S.,** Syntheses of isomaltose, isomaltotetraose, and isomaltooctaose, *Bull. Chem. Soc. Jpn.*, 46, 2520, 1973.
10. **Nicolaou, K. C., Dolle, R. E., Papahatjis, D. P., and Randall, J. L.,** Practical synthesis of oligosaccharides. Partial synthesis of avermectin B_{1a}, *J. Am. Chem. Soc.*, 106, 4189, 1984.
11. **Eby, R. and Schuerch, C.,** A chemical synthesis of benzylated methyl α-isomalto-oligosaccharides, *Macromolecules*, 7, 397, 1974.
12. **Eby, R. and Schuerch, C.,** The stepwise synthesis of methyl α-isomaltooligoside derivatives and methyl α-isomaltopentaoside, *Carbohydr. Res.*, 50, 203, 1976.
13. **Eby, R.,** The synthesis of α-isomalto-oligosaccharide derivatives and their protein conjugates, *Carbohydr. Res.*, 70, 75, 1979.
14. **Ogawa, T. and Takanashi, Y.,** Synthesis of β-D-(1→2)-linked D-glucopentaose, a part of the structure of the exocellular β-D-glucan of *Agrobacterium tumefaciens, Carbohydr. Res.*, 123, C16, 1983.
15. **Takeo, K. and Tei, S.,** Synthesis of the repeating units of Schizophyllan, *Carbohydr. Res.*, 145, 293, 1986.
16. **Takeo, K. and Suzuki, Y.,** Synthesis of the tri- and tetra-saccharides related to the fine structures of lichenan and cereal β-D-glucans, *Carbohydr. Res.*, 147, 265, 1986.
17. **Takeo, K., Okushio, K., Fukuyama, K., and Kuge, T.,** Synthesis of cellobiose, cellotriose, cellotetraose, and lactose, *Carbohydr. Res.*, 121, 163, 1983.
18. **Schmidt, R. R. and Michel, J.,** Synthese von linearen und verzweigten Cellotetraosen, *Angew. Chem.*, 94, 77, 1982.
19. **Nicolaou, K. C., Randall, J. L., and Furst, G. T.,** Stereospecific synthesis of rhynchosporosides: a family of fungal metabolites causing scald disease in barley and other grasses, *J. Am. Chem. Soc.*, 107, 5556, 1985.
20. **Ossowski, P., Pilotti, Å., Garegg, P. J., and Lindberg, B.,** Synthese eines verzweigten Hepta- und Octasaccharids mit Phytoalexin-Elicitor-Akitivität, *Angew. Chem.*, 95, 809, 1983.
21. **Ossowski, P., Pilotti, Å, Garegg, P. J., and Lindberg, B.,** Synthesis of a glucoheptaose and a glucooctaose that elicit phytoalexin accumulation in soybean, *J. Biol. Chem.*, 259, 11337, 1984.
22. **Takiura, K., Honda, S., Endo, T., and Kakehi, K.,** Studies of oligosaccharides. IX. Synthesis of gentio-oligosaccharides by block condensation, *Chem. Pharm. Bull.*, 20, 438, 1972.
23. **Excoffier, G., Gagnaire, D. Y., and Vignon, M. R.,** Le groupe trichloroacétyle comme substituant temporaire; synthése du gentiotétraose, *Carbohydr. Res.*, 46, 201, 1976.
24. **Takiura, K., Yamamoto, M., Miyaji, Y., Takai, H., Honda, S., and Yuki, H.,** Studies of oligosaccharides. XV. Syntheses of hydroquinone glycosides of gentio-oligosaccharides, *Chem. Pharm. Bull.*, 22, 2451, 1974.
25. **Excoffier, G., Paillet, M., and Vignon, M.,** Cyclic (1→6)-β-D-glucopyranose oligomers: synthesis of cyclogentiotriose and cyclogentiotetraose peracetates, *Carbohydr. Res.*, 135, C10, 1985.
26. **Klemer, A., Buhe, E., Gundlach, F., and Uhlemann, G.,** Synthesen und Abbau von Oligosacchariden, *Forschungsber. des Landes Nordheim-Westfalen*, Nr. 2393, 1974.
27. **Klemer, A. and Gundlach, F.,** Synthese eines verzweigten Tetrasaccharides: 6.6′-*Bis*-[β-D-glucosido ⟨1.5⟩]-maltose, *Chem. Ber.*, 96, 1765, 1963.
28. **Ogawa, T. and Kaburagi, T.,** Synthesis of a branched D-glucotetraose, the repeating unit of the extracellular polysaccharides of *Grifola umbellate, Sclerotinia libertiana, Porodisculus pendulus,* and *Schizophyllum commune* FRIES, *Carbohydr. Res.*, 103, 53, 1982.
29. **Kochetkov, N. K., Bochkov, A. F., Yazlovetskii, I. G., and Snyatkova, V. I.,** Synthesis of polysaccharides. I. Synthesis of galactoglucan of regular structure, *Izv. Akad. Nauk SSSR Ser. Khim.*, 1802, 1968.

30. **Chung, T. G., Ishihara, H., and Tejima, S.,** Synthesis of methyl *O*-β-D-galactopyranosyl-(1→6)-*O*-α-D-galactopyranosyl-(1→6)-*O*-α-D-galactopyranosyl-(1→6)-β-D-glucopyranoside, *Chem. Pharm. Bull.*, 27, 1589, 1979.
31. **Takamura, T., Chiba, T., Ishihara, H., and Tejima, S.,** Chemical modification of lactose. XIII. Synthesis of lacto-N-tetraose, *Chem. Pharm. Bull.*, 27, 1497, 1979.
32. **Takamura, T., Chiba, T., Ishihara, H., and Tejima, S.,** Chemical modification of lactose. XIII. Synthesis of lacto-*N*-tetraose, *Chem. Pharm. Bull.*, 28, 1804, 1980.
33. **Sato, S., Ito, Y., and Ogawa, T.,** Stereo- and regio-controlled, total synthesis of the Le^b antigen, $III^4FucIV^2FucLcOse_4Cer$, *Carbohydr. Res.*, 155, C1, 1986.
34. **Ponpipom, M. M., Bugianesi, R. L., and Shen, T. Y.,** Synthesis of paragloboside analogs, *Tetrahedron Lett.*, 1717, 1978.
35. **Zurabyan, S. E., Markin, V. A., Pimenova, V. V., Rozynov, B. V., Sadovskaya, V. L., and Khorlin, A. Ya.,** Synthesis of tri- and tetrasaccharides, structural isomers of milk oligosaccharides, *Bioorg. Khim.*, 4, 928, 1978.
36. **Paulsen, H., Paal, M., Hadamczyk, D., and Steiger, K.-M.,** Regioselective Glycosidierung von Lactose durch unterschiedliche Katalysatorsysteme. Synthese der Saccharid-Sequenzen von asialo-G_{M1}- und asialo-G_{M2}-Gangliosiden, *Carbohydr. Res.*, 131, C1, 1984.
37. **Maranduba, A. and Veyrières, A.,** Glycosylation of lactose. Synthesis of methyl *O*-(2-acetamido-2-deoxy-β-D-glucopyranosyl)-(1→3)-*O*-β-D-galactopyranosyl-(1→4)-β-D-glucopyranoside and methyl *O*-β-D-galactopyranosyl-(1→4)-*O*-(2-acetamido-2-deoxy-β-D-glucopyranosyl)-(1→3)-*O*-β-D-galactopyranosyl-(1→4)-β-D-glucopyranoside, *Carbohydr. Res.*, 135, 330, 1985.
38. **Dahmén, J., Gnosspelius, G., Larsson, A.-C., Lave, T., Noori, G., Pålsson, K., Frejd, T., and Magnusson, G.,** Synthesis of di-, tri-, and tetra-saccharides corresponding to receptor structures recognised by *Streptococcus pneumoniae, Carbohydr. Res.*, 138, 17, 1985.
39. **Maranduba, A. and Veyrières, A.,** Glycosylation of lactose: synthesis of branched oligosaccharides involved in the biosynthesis of glycolipids having blood-group I activity, *Carbohydr. Res.*, 151, 105, 1986.
40. **Takamura, T., Chiba, T., and Tejima, S.,** Chemical modification of lactose. XVI. Synthesis of lacto-*N*-neohexaose, *Chem. Pharm. Bull.*, 29, 587, 1981.
41. **Takamura, T., Chiba, T., and Tejima, S.,** Chemical modification of lactose. XVI. Synthesis of lacto-*N*-neohexaose, *Chem. Pharm. Bull.*, 29, 2270, 1981.
42. **Paulsen, H. and Bünsch, A.,** Synthese der Tetrasaccharid-Kette des P-Antigen-Globosids. Eine β-D-Glycosid-Synthese für 2-Amino-2-desoxyzucker, *Carbohydr. Res.*, 101, 21, 1982.
43. **Leontein, K., Nilsson, M., and Norberg, T.,** Synthesis of the methyl and 1-octyl glycosides of the P-antigen tetrasaccharide (globotetraose), *Carbohydr. Res.*, 144, 231, 1985.
44. **Paulsen, H. and Paal, M.,** Synthese der Tetra- und Trisaccharid-Sequenzen von Asialo-G_{M1} and G_{M2}. Lenkung der Regioselektivität der Glycosidierung von Lactose, *Carbohydr. Res.*, 137, 39, 1985.
45. **Sebesan, S. and Lemieux, R. U.,** Synthesis of tri- and tetrasaccharid haptens related to the asialo forms of the gangliosides G_{M2} and G_{M1}, *Can. J. Chem.*, 62, 644, 1984.
46. **Sugimoto, M., Horisaki, T., and Ogawa, T.,** Stereoselective synthesis of asialo-G_{M1}- and asialo-G_{M2}-ganglioside, *Glucoconjugate J.*, 2, 11, 1985.
47. **Takamura, T., Chiba, T., and Tejima, S.,** Chemical modification of lactose. XIV. Synthesis of *O*-2-acetamido-2-deoxy-β-D-glucopyranosyl-(1→3)-*O*-[2-acetamido-2-deoxy-β-D-glucopyranosyl-(1→6)]-*O*-β-D-galactopyranosyl-(1→4)-β-D-glucopyranose (3′,6′-di-β-*N*-acetylglucosaminyl-β-lactose), *Chem. Pharm. Bull.*, 29, 1027, 1981.
48. **Ito, Y., Sugimoto, S., and Ogawa, T.,** Total synthesis of a lacto-ganglio series glycosphingolipid, M1-XGL-1, *Tetrahedron Lett.*, 27, 4753, 1986.
49. **Sugimoto, M., Numata, M., Koike, K., Nakahara, Y., and Ogawa, T.,** Total synthesis of gangliosides G_{M1} and G_{M2}, *Carbohydr. Res.*, 156, C1, 1986.
50. **Kochetkov, N. K., Dmitriev, B. A., Nikolaev, A. V., Bayramova, N. E., and Shashkov, A. S.,** Synthesis of antigenic bacterial polysaccharides and their fragments. 10. Synthesis of hexasaccharide, a glucose analog of dimer of the repeating unit of a specific polysaccharide from *Salmonella newington, Bioorg. Khim.*, 5, 64, 1979.
51. **Paulsen, H. and Bünsch, H.,** Synthese der Pentasaccharid-Sequenz der Repeating-Unit der O-spezifischen Seitenkette des Lipopolysaccharides von *Shigella dysenteriae, Tetrahedron Lett.*, 22, 47, 1981.
52. **Paulsen, H. and Bünsch, H.,** Synthese der verzweigten Pentasaccharid-Einheit der O-spezifischen Seitenkette des Lipopolysaccharides von *Shigella dysenteriae, Chem. Ber.*, 114, 3126, 1981.
53. **Flowers, H. M.,** Synthesis of oligosaccharides of L-fucose containing α- and β-anomeric configurations in the same molecule, *Carbohydr. Res.*, 119, 75, 1983.
54. **Takeo, K. and Tei, S.,** Synthesis of lactodifucotetraose, *Carbohydr. Res.*, 141, 159, 1985.
55. **Fernandez-Mayoralas, A. and Martin-Lomas, M.,** Synthesis of 3- and 2′-fucosyl-lactose and 3,2′-difucosyllactose from partially benzylated lactose derivatives, *Carbohydr. Res.*, 154, 93, 1986.

56. **Morishima, N., Koto, S., Uchino, M., and Zen, S.,** Synthesis of a trifurcated tetrasaccharide using dehydrative glycosylation, *Chem. Lett.*, 1183, 1982.
57. **Paulsen, H. and Stenzel, W.,** Stereoselektive Synthese α-glycosidisch verknüpfter Di- and Oligosaccharide der 2-Amino-2-desoxy-D-glycopyranose, *Chem. Ber.*, 11, 2334, 1978.
58. **Zurabyan, S. E., Pimenova, V. V., Shashkova, E. A., and Khorlin, A. Ya.,** Synthesis of modified inhibitors of lysosim by oxazoline method, *Khim. Prir. Soedin.*, 7, 689, 1971.
59. **Warren, C. D., Jeanloz, R. W., and Strecker, G.,** Oligosaccharide oxazolines: preparation, and application to the synthesis of glycoprotein carbohydrate structures, *Carbohydr. Res.*, 92, 85, 1981.
60. **Warren, C. D., Milat, M.-L., Augé, C., and Jeanloz, R. W.,** The synthesis of a trisaccharide and a tetrasaccharide lipid intermediate. P^1-Dolichyl P^2-[*O*-β-D-mannopyranosyl-(1→4)-*O*-(2-acetamido-2-deoxy-β-D-glucopyranosyl-(1→4)-2-acetamido-2-deoxy-α-D-glycopyranosyl]diphosphate and P^1 dolichyl P^2-[*O*-α-D-mannopyranosyl-(1→3)-*O*-β-D-mannopyranosyl-(1→4)-*O*-(2-acetamido-2-deoxy-β-D-glucopyranosyl-(1→4)-2-acetamido-2-deoxy-α-D-glycopyranosyl]diphosphate, *Carbohydr. Res.*, 126, 61, 1984.
61. **Warren, C. D., Jeanloz, R. W., and Strecker, G.,** The synthesis of tri- and tetra-saccharide oxazolines, *Carbohydr. Res.*, 71, C5, 1979.
62. **Paulsen, H., Stiem, M., and Unger, F. M.,** Synthese eines 3-Desoxy-D-manno-2-octulosonsäure (KDO)-haltigen Tetrasaccharides und dessen Strukturvergleich mit einem Abbauprodukt aus bakterien-Lipopolysacchariden, *Tetrahedron Lett.*, 27, 1135, 1986.
63. **Itoh, Y. and Tejima, S.,** Syntheses of acetylated tetrasaccharides, Manα1→2-Manα1→3Manβ1→4GlcNAc and Manα1→3Manα1→6Manβ1→4GlcNAc, *Chem. Pharm. Bull.*, 32, 957, 1984.
64. **Paulsen, H., Lebuhn, R., and Lockhoff, O.,** Synthese des verzweigten Tetrasaccharid-Bausteins der Schlüsselsequenz von N-Glycoproteinen, *Carbohydr. Res.*, 103, C7, 1982.
65. **Paulsen, H. and Lebuhn, R.,** Synthese von Tri- und Tetrasaccharid-Sequenzen von *N*-Glycoproteinen mit β-D-mannosidischer Verknüpfung, *Liebigs Ann. Chem.*, 1047, 1983.
66. **Bundle, D. R. and Josephson, S.,** Artificial carbohydrate antigens: the synthesis of a tetrasaccharide hapten, a *Shigella flexneri* O-antigen repeating unit, *Carbohydr. Res.*, 80, 75, 1980.
67. **Beetz, T. and van Boeckel, C. A. A.,** Synthesis of an antithrombin binding heparin-like pentasaccharide, *Tetrahedron Lett.*, 27, 5889, 1986.
68. **Sinaÿ, P., Jacquinet, J.-C., Petitou, M., Duchaussoy, P., Lederman, I., Choay, J., and Torri, G.,** Total synthesis of a heparin pentasaccharide fragment having high affinity for antithrombin III, *Carbohydr. Res.*, 132, C5, 1984.
69. **Petitou, M., Duchaussoy, P., Lederman, I., Choay, J., Sinaÿ, P., Jacquinet, J.-C., and Torri, G.,** Synthesis of heparin fragments. A chemical synthesis of the pentasaccharide *O*-(2-deoxy-2-sulfamido-6-*O*-sulfo-α-D-glucopyranosyl)-(1→4)-*O*-(β-D-glucopyranosyluronic acid-(1→4)-*O*-(2-deoxy-2-sulfamido-3,6-di-*O*-sulfo-α-D-glucopyranosyl)-(1→4)-*O*-(2-*O*-sulfo-α-L-idopyranosyluronic acid)-(1→4)-2-deoxy-2-sulfamido-6-*O*-sulfo-D-glucopyranose decasodium salt, a heparin fragment having high affinity for antithrombin III, *Carbohydr. Res.*, 147, 221, 1986.
70. **van Boeckel, C. A. A., Beetz, T., Vos, J. N., de Jong, A. J. M., van Aelst, S. F., van den Bosch, R. H., Mertens, J. M. R., and van der Vlugt, F. A.,** Synthesis of a pentasaccharide corresponding to the antithrombin III binding fragment of heparin, *J. Carbohydr. Chem.*, 4, 293, 1985.
71. **Ichikawa, Y., Monden, R., and Kuzuhara, H.,** Synthesis of a haparin pentasaccharide fragment with a high affinity for antithrombin III employing cellobiose as a key starting material, *Tetrahedron Lett.*, 27, 611, 1986.
72. **Paulsen, H. and Kolář, Č.,** Synthese der Tetrasaccharidketten der Determinanten der Blutgruppensubstanzen A and B, *Angew. Chem.*, 90, 823, 1978.
73. **Paulsen, H. and Kolář, Č.,** Synthese der Oligosaccharid-Determinanten der Blutgruppensubstanzen der Type 1 des ABH-Systems, *Chem. Ber.*, 112, 3190, 1979.
74. **Bovin, N. N. and Khorlin, A. Ya.,** Synthesis of the determinant oligosaccharides of ABH (type 1) blood group antigens and Le^b tetrasaccharide from the same precursor, *Bioorg. Khim.*, 10, 853, 1984.
75. **Nashed, M. A. and Anderson, L.,** Oligosaccharides from "standardized intermediates". Synthesis of a branched tetrasaccharide glycoside related to the blood group B determinant, *J. Am. Chem. Soc.*, 104, 7282, 1982.
76. **Bovin, N. V., Zurabyan, S. E., and Khorlin, A. Ya.,** Stereoselectivity of glycosylation with derivatives of 2-azido-2-deoxy-D-galactopyranose. The synthesis of a determinant oligosaccharide related to blood-group A (Type 1), *Carbohydr. Res.*, 112, 23, 1983.
77. **Bovin, N. V., Zurabyan, S. E., and Khorlin, A. Ya.,** Stereospecific glycosylation with 2-azido-2-deoxy-D-galactopyranose derivatives and synthesis of determinant oligosaccharide related to the blood-group A, type 1, *Izv. Akad. Nauk SSSR Ser. Khim.*, 1148, 1982.
78. **Bovin, N. V., Zurabyan, S. E., and Khorlin, A. Ya.,** Synthesis of blood-group A determinant oligosaccharides, *Bioorg. Khim.*, 7, 1271, 1981.
79. **Veyrières, A.,** Blood-group Ii-active oligosaccharides. Synthesis of a tetrasaccharide, a β-(1→3) dimer of *N*-acetyl-lactosamine, *J. Chem. Soc. Perkin Trans. I*, 1626, 1981.

80. **Alais, J. and Veyrières, A.,** Block synthesis of a hexasaccharide hapten of i blood group antigen, *Tetrahedron Lett.*, 24, 5223, 1983.
81. **Milat, M.-L. and Sinaÿ, P.,** Synthesis of the tetrasaccharide *O*-α-L-fucopyranosyl-(1→2)-*O*-α-D-galactopyranosyl-(1→3)-*O*-β-D-galactopyranosyl-(1→4)-2-acetamido-2-deoxy-D-glucopyranose, the antigenic determinant of human blood-group B (Type 2), *Carbohydr. Res.*, 92, 183, 1981.
82. **Milat, M.-L., and Sinaÿ, P.,** Synthese der antigenen Determinante der Blutgruppensubstanz B: Das Tetrasaccharid vom Typ 2, *Angew. Chem.*, 92, 501, 1979.
83. **Paulsen, H. and Kolář, Č.,** Synthese der Tetrasaccharid-Ketten der Type 2 der Determinanten der Blutgruppensubstanzan A und B, *Tetrahedron Lett.*, 2881, 1979.
84. **Paulsen, H. and Kolář, Č.,** Synthese der Oligosaccharid-Determinanten der Blutgruppensubstanzes der Type 2 des ABH-Systems. Diskussion der α-Glycosid-Synthese, *Chem. Ber.*, 114, 306, 1981.
85. **Nashed, M. A. and Anderson, L.,** Oligosaccharides from "standardized intermediates". Synthesis of a branched tetrasaccharide glycoside isomeric with the blood-group B, Type 2 determinant, *Carbohydr. Res.*, 114, 43, 1983.
86. **Amvam-Zollo, P.-H., and Sinaÿ, P.,** *Streptococcus pneumoniae* Type XIV polysaccharide: synthesis of a repeating branched tetrasaccharide with dioxa-type spacer-arms, *Carbohydr. Res.*, 150, 199, 1986.
87. **Zurabyan, S. E., Nesmeyanov, V. A., and Khorlin, A. Ya.,** Synthesis of the branched tetrasaccharide 2-acetamido-4-*O*-β-D-galactopyranosyl-6-*O*-[*O*-β-D-galactopyranosyl-(1→4)-β-D-glucopyranosyl]-2-deoxy-D-glucose, *Izv. Akad. Nauk SSSR Ser. Khim.*, 1421, 1976.
88. **Rana, S. S., Barlow, J. J., and Matta, K. L.,** The chemical synthesis of *O*-α-L-fucopyranosyl-(1→2)-*O*-β-D-galactopyranosyl-(1→3)-*O*-[α-L-fucopyranosyl-(1→4)]-2-acetamido-2-deoxy-D-glucopyranose, the Lewis b blood-group antigenic determinant, *Carbohydr. Res.*, 96, 231, 1981.
89. **Hindsgaul, O., Norberg, T., Le Pendu, J., and Lemieux, R. U.,** Synthesis of type 2 human blood-group antigenic determinants. The H, X, and Y haptens and variations of the H type 2 determinant as probes for the combining site of the lectin I of *Ulex europaeus, Carbohydr. Res.*, 109, 109, 1982.
90. **Jacquinet, J.-C. and Sinaÿ, P.,** Synthesis of blood-group substances. 6. Synthesis of O-α-L-fucopyranosyl-(1→2)-*O*-β-D-galactopyranosyl-(1→4)-*O*-[α-L-fucopyranosyl-(1→3)]-2-acetamido-2-deoxy-α-D-glucopyranose, the postulated Lewis d antigenic determinant, *J. Org. Chem.*, 42, 720, 1977.
91. **Ogawa, T., Nukada, T., and Kitajima, T.,** Synthesis of a linear, hexahexosyl unit of a high-mannose type of glycan chain of a glycoprotein, *Carbohydr. Res.*, 123, C12, 1983.
92. **Kitajima, T., Sugimoto, M., Nukada, T., and Ogawa, T.,** Synthesis of a linear tetrasaccharide unit of a complex type of glycan chain of a glycoprotein, *Carbohydr. Res.*, 127, C1, 1984.
93. **Paulsen, H. and Tietz, H.,** Synthese eines *N*-Acetylneuraminsäure-haltigen Syntheseblocks. Kupplung zum *N*-Acetylneuraminsäure-Tetrasaccharid mit Trimethylsilyltriflat, *Carbohydr. Res.*, 144, 205, 1985.
94. **Paulsen, H. and Tietz, H.,** Herstellung eines *N*-Acetylneureminsäure-haltigen Trisaccharids und dessen Verwendung in Oligosaccharidsynthesen, *Angew. Chem.*, 97, 118, 1985.
95. **Ogawa, T., Sugimoto, M., Kitajima, T., Sadozai, K. K., and Nukada, T.,** Total synthesis of a undecasaccharide: a typical carbohydrate sequence for the complex type of glycan chains of a glycoprotein, *Tetrahedron Lett.*, 27, 5739, 1986.
96. **Lönn, H.,** Synthesis of a tetra- and a nona-saccharide which contain α-L-fucopyranosyl groups and are part of the complex type of carbohydrate moiety of glycoproteins, *Carbohydr. Res.*, 139, 115, 1985.
97. **Ogawa, T. and Yamamoto, H.,** Synthesis of linear D-mannotetraose and D-mannohexaose, partial structures of the cell-surface D-mannan of *Candida albicans* and *Candida utilis, Carbohydr. Res.*, 104, 271, 1982.
98. **Ogawa, T. and Yamamoto, H.,** Synthesis of a model, linear D-mannopentaose for the nonreducing-end sequence of the cell-surface D-mannan of *Escherichia coli, Candida albicans,* and *Saccharomyces cerevisiae, Carbohydr. Res.*, 137, 79, 1985.
99. **Ogawa, T. and Yamamoto, H.,** Synthesis of model linear mannohexaose for the backbone structure of fruit body polysaccharide of *Tremella fuciformis* and *Dictiophora indusiata* FISCH, *Agric. Biol. Chem.*, 49, 475, 1985.
100. **Ogawa, T. and Nukada, T.,** Synthesis of a branched mannohexaoside, a part structure of a high-mannose-type glycan of a glycoprotein, *Carbohydr. Res.*, 136, 135, 1985.
101. **Shah, R. N., Cumming, D. A., Grey, A. A., Carver, J. P., and Krepinsky, J. J.,** Synthesis of a tetrasaccharide of the extended core-region of the saccharide moiety of N-linked glycoproteins, *Carbohydr. Res.*, 153, 155, 1986.
102. **Tahir, S. H. and Hindsgaul, O.,** Substrates for the differentiation of the *N*-acetylglucosaminyl transferases. Synthesis of βDGlc*N*Ac(1→2)αDMan(1→6)βDMan and βDGlcNAc(1→2)αDMan(1→6) αDMan(1→3) βDMan glycosides, *Can. J. Chem.*, 64, 1771, 1986.
103. **Ogawa, T., Katano, K., Sasajima, K., and Matsui, M.,** Synthetic studies on cell surface glycans. 3. Branching pentasaccharides of glycoprotein, *Tetrahedron*, 37, 2779, 1981.
104. **Winnik, F. M., Carver, J. P., and Krepinsky, J. J.,** Synthese of model oligosaccharides of biological significance. 2. Synthesis of a tetramannoside and of two lyxose-containing trisaccharides, *J. Org. Chem.*, 47, 2701, 1982.

105. **Arnarp, J., Haraldsson, M., and Lönngren, J.,** Synthesis of three oligosaccharides that form part of the complex type of carbohydrate moiety of glycoproteins containing intersecting *N*-acetylglucosamine, *J. Chem. Soc. Perkin Trans. I,* 535, 1985.
106. **Takeda, T., Kawarasaki, I., and Ogihara, Y.,** Studies on the structure of a polysaccharide from *Epidermophyton floccosum* and approach to a synthesis of the basic trisaccharide repeating units, *Carbohydr. Res.,* 89, 301, 1981.
107. **Paulsen, H. and Lorentzen, J. P.,** Synthese von Tri- und Tetrasaccharid-Einheit des *O*-Antigens aus *Aeromonas salmonicida, Tetrahedron Lett.,* 26, 6043, 1985.
108. **Wessel, H.-P. and Bundle, D. R.,** Artificial carbohydrate antigens: a block synthesis of a linear, tetrasaccharide repeating-unit of the *Shigella flexneri* variant Y polysaccharide, *Carbohydr. Res.,* 124, 301, 1983.
109. **Bock, K. and Meldal, M.,** Synthesis of tetrasaccharides related to the O-specific determinants of *Salmonella* serogroups A, B and D_1, *Acta Chem. Scand. B,* 38, 255, 1984.
110. **Bock, K. and Meldal, M.,** Synthesis of the branchpoint tetrasaccharide of the O-specific determinant of Salmonella serogroup B, *Acta Chem. Scand. B,* 38, 71, 1984.
111. **Iversen, T., Josephson, S., and Bundle, D. R.,** Synthesis of *Streptococcal groups A, C and variant-A antigenic determinants, J. Chem. Soc. Perkin. Trans. I,* 2379, 1981.
112. **Pozsgay, V., Nánási, P., and Neszmélyi, A.,** Approach to the synthesis of the repeating units of immunodeterminant bacterial polysaccharides: synthesis and ^{13}C N. M. R. analysis of β-D-galactopyranosyl-(1→3)-α-L-rhamnopyranosyl-(1→3)-α-L-rhamnopyranosyl-(1→2)-L-rhamnopyranose, *J. Chem. Soc. Chem. Commun.,* 828, 1979.
113. **Lipták, A., Szabó, L., Kerékgyártó, J., Harangi, J., Nánási, P., and Duddeck, H.,** Synthesis of the tetrasaccharide repeating-unit of the lipopolysaccharide isolated from *Pseudomonas maltophilia, Carbohydr. Res.,* 150, 187, 1986.
114. **Backinowsky, L. V., Gomtsyan, A. R., Byramova, N. E., and Kochetkov, N. K.,** Synthesis of oligosaccharide fragments of *Shigella flexneri* O-specific polysaccharide. II. Synthesis of trisaccharide Glcα1→3Rhaα1→2Rhaα1→OMe and tetrasaccharide GlcNAcβ1→2(Glcα1→3)Rhaα1→2Rhaα1→OMe, *Bioorg. Khim.,* 11, 254, 1985.
115. **Josephson, S. and Bundle, D. R.,** Artificial carbohydrate antigens: the synthesis of the tetrasaccharide repeating unit of *Shigella flexneri* O antigen, *Can. J. Chem.,* 57, 3073, 1979.
116. **Tsvetkov, Yu. E., Byramova, N. E., Backinowsky, L. V., Kochetkov, N. K., and Yankina, N. F.,** Synthesis of a repeating unit of the basic chain of *Shigella flexneri* O-antigenic polysaccharides, *Bioorg. Khim.,* 12, 1213, 1986.
117. **Byramova, N. É., Tsvetkov, Yu. E., Backinowsky, L. V., and Kochetkov, N. K.,** Synthesis of the basic chain of the *O*-specific polysaccharides of *Shigella flexneri, Carbohydr. Res.,* 137, C8, 1985.
118. **Kochetkov, N. K., Byramova, N. É., Tsvetkov, Yu. E., and Backinowsky, L. V.,** Synthesis of the O-specific polysaccharide of *Shigella flexneri, Tetrahedron,* 41, 3363, 1985.
119. **Byramova, N. É., Tsvetkov, Yu. E., Backinowsky, L. V., and Kochetkov, N. K.,** Synthesis of the O-specific polysaccharide of *Shigella flexneri.* 5. Synthesis of *O*-(4,6-di-*O*-benzoyl-2-deoxy-2-phthaloimido-β-D-glucopyranosyl)-(1→2)-*O*-(3,4-di-*O*-benzoyl-α-L-rhamnopyranosyl)-(1→2)-*O*-(3,4-di-*O*-benzoyl-α-L-rhamnopyranosyl)-(1→3)-4-*O*-benzoyl-1,2-*O*-[1-(*exo*-cyano)ethylidene]-β-L-rhamnopyranose suitable monomer for polycondensation, *Izv. Akad. Nauk SSSR Ser. Khim.,* 1145, 1985.
120. **Backinowsky, L. V., Gomtsyan, A. R., Byramova, N. É., Kochetkov, N. K., and Yankina, N. F.,** Synthesis of oligosaccharide fragments of *Shigella flexneri* O-specific polysaccharides. IV. The synthesis of the trisaccharide Glcα1→3Rhaα1→3Rhaα-OMe and tetrasaccharides Rhaα1→2(Glcα1→3)Rhaα1→3Rhaα-OMe and GlcNAcβ1→2Rhaα1→2(Glcα1→3)Rhaα-OMe. Localization of the O-factor V, *Bioorg. Khim.,* 11, 1562, 1985.
121. **Paulsen, H. and Kutschker, W.,** Synthese von β-L-rhamnosidish verknüpften Oligosacchariden des Lipopolysaccharides aus *Shigella flexneri* Serotyp 6, *Carbohydr. Res.,* 120, 25, 1983.
122. **Paulsen, H. and Kutschker, W.,** Synthese einer verzweigten Tetrasaccharid-Einheit der O-spezifischen Kette des Lipopolysaccharide aus *Shigella flexneri* Serotyp 6, *Liebigs Ann. Chem.,* 557, 1983.
123. **Paulsen, H. and Lockhoff, O.,** Neue effektive β-Glycosidsynthese für Mannose-Glycoside Synthesen von Mannose-haltigen Oligosacchariden, *Chem. Ber.,* 114, 3102, 1981.
124. **Paulsen, H. and Lockhoff, O.,** Synthese der Repeating-Unit der O-spezifischen Kette des Lipopolysaccharides des Bakteriums *Escherichia coli* O 75, *Chem. Ber.,* 114, 3115, 1981.
125. **Schwarzenbach, D. and Jeanloz, R. W.,** Synthesis of part of the antigenic repeating-unit of *Streptococcus pneumoniae* type II, *Carbohydr. Res.,* 90, 193, 1981.
126. **Gomtsyan, A. R., Byramova, N. É., Backinowsky, L. B., and Kochetkov, N. K.,** Synthesis of a branched pentasaccharide fragment of the O-antigen of *Shigella flexneri* serotype 5b, *Carbohydr. Res.,* 138, C1, 1985.

127. **Backinowsky, L. V., Gomtsyan, A. R., Byramova, N. É., and Kochetkov, N. K.,** Synthesis of oligosaccharide fragments of *Shigella flexneri* O-specific polysaccharides. III. Synthesis of the tetrasaccharide Glcα1→3Rhaα1→2(Glcα1→3)Rhaα-1-OMe and pentasaccharide GlcNAcβ1→2(Glcα1→3)Rhaα1→2(Glcα1→3)Rhaα1-OMe, *Bioorg. Khim.*, 11, 655, 1985.
128. **Wessel, H.-P. and Bundle, D. R.,** Strategies for the synthesis of branched oligosaccharides of the *Shigella flexneri* 5a, 5b, and variant X serogroups employing a multifunctional rhamnose precursor, *J. Chem. Soc. Perkin Trans. I*, 2251, 1985.
129. **Paulsen, H. and Lorentzen, J. P.,** Synthese von verzweigten Tri- und Tetrasaccharidsequenzen der "Repeating Unit" der O-spezifischen Kette des Lipopolysaccharides aus *Aeromonas salmonicida, Liebigs Ann. Chem.*, 1586, 1986.
130. **Takeo, K., Nakaji, T., and Shinmitsu, K.,** Synthesis of lycotetraose, *Carbohydr. Res.*, 133, 275, 1984.
131. **Garegg, P. J. and Norberg, T.,** Synthesis of the biological repeating units of Salmonella serogroups A, B, and D_1 O-antigenic polysaccharides, *J. Chem. Soc. Perkin Trans. I*, 2973, 1982.
132. **Kováč, P., Taylor, R. B., and Glaudemans, C. P. J.,** General synthesis of (1→3)-β-D-galacto oligosaccharides and their methyl β-glycosides by a stepwise or a blockwise approach, *J. Org. Chem.*, 50, 5323, 1985.
133. **Koeners, H. J., Verhoeven, J., and van Boom, J. H.,** Application of levulinic acid ester as a protective groups in the synthesis of oligosaccharides, *Recl. Trav. Chim. Pays-Bas*, 100, 65, 1981.
134. **Nashed, E. M. and Glaudemans, C. P. J.,** Synthesis of 2,3-epoxypropyl β-glycosides of (1→6)-β-D-galactopyranooligosaccharides and their binding to monoclonal anti-galactan IgA J539 and IgA X24, *Carbohydr. Res.*, 158, 125, 1986.
135. **Srivastava, V. K. and Schuerch, C.,** Synthesis and characterization of a tetramer, propyl 6-*O*-β-D-galactopyranosyl-(1→6)-*O*-β-D-galactopyranosyl-(1→6)-*O*-β-D-galactopyranosyl-(1→6)-β-D-galactopyranoside, *Carbohydr. Res.*, 106, 217, 1982.
136. **Kováč, P.,** Efficient chemical synthesis of methyl β-glycosides of β-(1→6)-linked D-galacto-oligosaccharides by a stepwise and a blockwise approach, *Carbohydr. Res.*, 153, 237, 1986.
137. **Kováč, P.,** Systematic chemical synthesis of (1→6)-β-D-galacto-oligosaccharides and related compounds, *Carbohydr. Res.*, 144, C12, 1985.
138. **Kochetkov, N. K., Torgov, V. I., Malysheva, N. N., Shashkov, A. S., and Klimov, E. M.,** Synthesis of the tetrasaccharide repeating unit of the O-specific polysaccharide from *Salmonella muenster* and *Salmonella minneapolis, Tetrahedron*, 36, 1227, 1980.
139. **Kochetkov, N. K., Malysheva, N. N., Torgov, V. I., and Klimov, E. M.,** Synthesis of antigenic bacterial polysaccharides and their fragments. 4. Synthesis of an analog of a biological repeating unit of O-antigenic polysaccharide from *Salmonella senftenberg, Izv. Akad. Nauk SSSR Ser. Khim.*, 654, 1977.
140. **Kochetkov, N. K., Malysheva, N. N., Torgov, V. I., and Klimov, E. M.,** Synthesis of the tetrasaccharide repeating-unit of the O-specific polysaccharide from *Salmonella senftenberg, Carbohydr. Res.*, 54, 269, 1977.
141. **Koeners, H. J., Verdegaal, C. H. M., and van Boom, J. H.,** 4,4-(Ethylenedithio)pentanoyl: a masked levulinoyl protective group in the synthesis of oligosaccharides, *Recl. Trav. Chim. Pays-Bas*, 100, 118, 1981.
142. **Augé, C., David, S., and Veyrières, A.,** Synthesis of a branched pentasaccharide: one of the core oligosaccharides of human blood-group substances, *J. Chem. Soc. Chem. Commun.*, 449, 1977.
143. **Augé, C., David, S., and Veyrières, A.,** Molecular basis of the human Ii blood-group systems. Contribution to the problem from the synthesis of I-active oligosaccharides, *Nouv. J. Chim.*, 3, 491, 1979.
144. **Schmidt, R. R. and Grundler, G.,** Synthese eines Tetrasaccharids der Core-Region von O-Glycoproteinen, *Angew. Chem.*, 95, 805, 1983.
145. **Grundler, G. and Schmidt, R. R.,** Anwendung des Trichloracetimidatverfahrens auf 2-Desoxy-2-phthalimido-D-glucose-Derivate. Synthese von Oligosacchariden der "Core-Region" von *O*-Glycoproteinen des Mucin-Typs, *Carbohydr. Res.*, 135, 203, 1985.
146. **Piskorz, C. F., Abbas, S. A., and Matta, K. L.,** Synthetic mucin fragments: benzyl 2-acetamido-6-*O*-(2-acetamido-2-deoxy-β-D-glucopyranosyl)-2-deoxy-3-*O*-β-D-galactopyranosyl-α-D-galactopyranoside and benzyl 2-acetamido-6-*O*-(2-acetamido-2-deoxy-β-D-glucopyranosyl)-3-*O*-[6-*O*-(2-acetamido-2-deoxy-β-D-glucopyranosyl)-β-D-galactopyranosyl]-2-deoxy-α-D-galactopyranoside, *Carbohydr. Res.*, 126, 115, 1984.
147. **Kováč, P. and Glaudemans, C. P. J.,** Synthesis of methyl glycosides of β-(1→6)-linked D-galactobiose, galactoriose, and galactotetraose having a 3-deoxy-3-fluoro-β-D-galactopyranoside end-residue, *Carbohydr. Res.*, 140, 289, 1985.
148. **Takeo, K., Aspinall, G. O., Brennan, P. J., and Chatterjee, D.,** Synthesis of tetrasaccharides related to the antigenic determinants from the glycopeptidolipid antigens of serovars 9 and 25 in the *Mycobacterium avium-M. intracellulare-M. scrofulaceum* serocomplex, *Carbohydr. Res.*, 150, 133, 1986.
149. **Thiem, J. and Ossowski, P.,** Synthesen von Digitoxosyl-Digitoxosiden und Herzglycosid-Tetrasacchariden, *Liebigs Ann. Chem.*, 2215, 1983.

150. **Suami, T., Otake, T., Nishiyama, S., Adachi, R., and Nakamura, K.**, Synthesis of stachyose tetradecaacetate and an isomer, *Bull. Chem. Soc. Jpn.*, 51, 1826, 1978.
151. **Kováč, P., Hirsch, J., Kováčik, V. and Kočiš, P.**, The stepwise synthesis of an aldopentaouronic acid derivative related to branched (4-*O*-methylglucurono)xylans, *Carbohydr. Res.*, 85, 41, 1980.
152. **Sakai, J., Sawaki, M., and Takeda, T.**, Synthesis of partial structural unit of lichenan, *Nippon Kagaku Kaishi*, 1657, 1982.
153. **Paulsen, H. and Bünsch, A.**, Synthese der Pentasaccharidkette des Forssman-Antigens, *Angew. Chem.*, 92, 929, 1980.
154. **Paulsen, H. and Bünsch, A.**, Reaktivitätsuntersuchungen bei Tri- und Pentasaccharidsynthesen. Verbesserte Synthese der Pentasaccharidkette des Forssman-Antigens, *Liebigs Ann. Chem.*, 2204, 1981.
155. **Paulsen, H. and Bünsch, A.**, Synthese der Pentasaccharid-Kette des Forssman-Antigens, *Carbohydr. Res.*, 100, 143, 1982.
156. **Paulsen, H. and Lebuhn, R.**, Synthese der invarianten Pentasaccharid-Core-Region der Kohlenhydrat-Ketten der N-Glycoproteine, *Carbohydr. Res.*, 130, 85, 1984.
157. **Paulsen, H. and Lebuhn, R.**, Synthese von Pentasaccharid- und Octasaccharid-Sequenzen der Kohlenhydrat-Kette von *N*-Glycoproteinen, *Carbohydr. Res.*, 125, 21, 1984.
158. **Paulsen, H. and Lebuhn, R.**, Synthese eines Octasaccharids der Basissequenz von N-Glycoproteinen, *Angew. Chem.*, 94, 933, 1982.
159. **Paulsen, H., Rauwald, W., and Lebuhn, R.**, Synthese von unsymmetrischen Pentasaccharid-Sequenzen der N-Glycoproteine, *Carbohydr. Res.*, 138, 29, 1985.
160. **Paulsen, H., Heume, M., Györgydeák, Z., and Lebuhn, R.**, Synthese einer verzweigten Pentasaccharid-Sequenz der "bisected" Struktur von *N*-Glycoproteinen, *Carbohydr. Res.*, 144, 57, 1985.
161. **Arnarp, J. and Lönngren, J.**, Synthesis of hepta- and penta-saccharides, part of the complex-type carbohydrate portion of glycoproteins, *J. Chem. Soc. Chem. Commun.*, 1000, 1980.
162. **Arnarp, J. and Lönngren, J.**, Synthesis of a tri-, a penta-, and a hepta-saccharide containing terminal *N*-acetyl-β-D-lactosaminyl residues, part of the "complex-type" carbohydrate moiety of glycoproteins, *J. Chem. Soc. Perkin Trans. I*, 2070, 1981.
163. **Arnarp, J., Baumann, H., Lönn, H., Lönngren, J., Nyman, H., and Ottoson, H.**, Synthesis of oligosaccharides that form parts of the complex type of carbohydrate moieties of glycoproteins. Three glycosides prepared for conjugation to proteins, *Acta Chem. Scand. B*, 37, 329, 1983.
164. **Sadozai, K. K., Nukada, T., Ito, Y., Nakahara, Y., Ogawa, T., and Kobata, A.**, Synthesis of a heptasaccharide hapten related to a biantennary glycan chain of human chorionic gonadotropin of a choriocarcinoma patient. A convergent approach, *Carbohydr. Res.*, 157, 101, 1986.
165. **Arnarp, J., Haraldsson, M., and Lönngren, J.**, Synthesis of three oligosaccharides that form part of the complex type of carbohydrate moiety of glycoproteins, *Carbohydr. Res.*, 97, 307, 1981.
166. **Ogawa, T. and Sasajima, K.**, Synthesis of a model of an inner chain of cell-wall proteoheteroglycan isolated from *Piricularia oryzae*: branched D-mannopentaosides, *Carbohydr. Res.*, 93, 67, 1981.
167. **Sadozai, K. K., Kitajima, T., Nakahara, Y., Ogawa, T., and Kobata, A.**, Synthesis of a pentasaccharide hapten related to a monoantennary glycan chain of human chorionic gonadotropin, *Carbohydr. Res.*, 152, 173, 1986.
168. **Sadozai, K. K., Ito, Y., Nukada, T., Ogawa, T., and Kobata, A.**, Synthesis of a heptasaccharide hapten related to an anomalous biantennary glycan-chain of human chorionic gonadotropin of a patient with choriocarcinoma. A stepwise approach, *Carbohydr. Res.*, 150, 91, 1986.
169. **Sadozai, K. K., Nukada, T., Ito, Y., Kobata, A., and Ogawa, T.**, Synthesis of a heptasaccharide hapten related to an anomalous biantennary glycan chain of human chorionic gonadotropin of a patient with a choriocarcinoma, *Agric. Biol. Chem.*, 50, 251, 1986.
170. **Ogawa, T., Katano, K., and Matsui, M.**, Regio- and stereo-controlled synthesis of core oligosaccharides of glycopeptides, *Carbohydr. Res.*, 64, C3, 1978.
171. **Ogawa, T. and Sasajima, K.**, Synthesis of a branched D-mannopentaoside and a branched D-mannohexaoside: models of the inner core of cell-wall glycoproteins of *Saccharomyces cerevisiae, Carbohydr. Res.*, 93, 231, 1981.
172. **Ogawa, T. and Sasajima, K.**, Synthesis of a branched D-mannopentaoside and a branched D-mannohexaoside: models of the outer chain of the glycan of soybean agglutinin, *Carbohydr. Res.*, 93, 53, 1981.
173. **Ogawa, T. and Sasajima, K.**, Reconstruction of glycan chains of glycoprotein. Branching mannopentaoside and mannohexaoside, *Tetrahedron*, 37, 2787, 1981.
174. **Kochetkov, N. K., Torgov, V. I., Malysheva, N. N., and Shashkov, A. S.**, Synthesis of the pentasaccharide repeating unit of the O-specific polysaccharide from *Salmonella strasbourg, Tetrahetron*, 36, 1099, 1980.
175. **Baumann, H., Lönn, H., and Lönngren, J.**, Synthesis of a hexasaccharide that forms part of an alveolar glycoprotein, *Carbohydr. Res.*, 114, 317, 1983.
176. **Ogawa, T. and Nakabayashi, S.**, Synthesis of a hexasaccharide unit of a complex type of glycan chain of a glycoprotein, *Carbohydr. Res.*, 93, C1, 1981.

177. **Ogawa, T., Nakabayashi, S., and Kitajima, T.,** Synthesis of a hexasaccharide unit of a complex type of glycan chain of a glycoprotein, *Carbohydr. Res.*, 114, 225, 1983.
178. **Kochetkov, N. K., Nikolaev, A. V., and Dmitriev, B. A.,** Synthesis of of hexa- and nonasaccharide units of O-antigenic polysaccharide of *Salmonella newington, Izv. Akad. Nauk SSSR Ser. Khim.*, 699, 1981.
179. **Kochetkov, N. K.,** Ways of synthesizing O-antigenic polysaccharides, *Izv. Akad. Nauk SSSR Ser. Khim.*, 243, 1984.
180. **Dmitriev, B. A., Nikolaev, A. V., Shashkov, A. S., and Kochetkov, N. K.,** Block-synthesis of higher oligosaccharides: synthesis of hexa- and nonsaccharide fragments of the O-antigenic polysaccharide of *Salmonella newington, Carbohydr. Res.*, 100, 195, 1982.
181. **Ogawa, T. and Kaburagi, T.,** Synthesis of a branched D-glycoheptaose: the repeating unit of extracellular α-D-glucan 1355-S *Leuconostoc mesenteroides* NRRLB-1355, *Carbohydr. Res.*, 110, C12, 1982.
182. **Lönn, H.,** Synthesis of a tri- and a hepta-saccharide which contain α-L-fucopyranosyl groups and are part of the complex type of carbohydrate moiety of glycoproteins, *Carbohydr. Res.*, 139, 105, 1985.
183. **Ogawa, T., Kitajima, T., and Nukada, T.,** Synthesis of a nonahexosyl unit of a complex type of glycan chain of a glycoprotein, *Carbohydr. Res.*, 123, C8, 1983.
184. **Arnarp, J., Haraldsson, M., and Lönngren, J.,** Synthesis of a nonasaccharide containing terminal *N*-acetyl-β-D-lactosaminyl residues, part of the complex-type carbohydrate moiety of glycoproteins, *J. Chem. Soc. Perkin Trans. I,* 1841, 1982.
185. **Lönn, H. and Lönngren, J.,** Synthesis of a nona- and an undeca-saccharide that form part of the complex type of carbohydrate moiety of glycoproteins, *Carbohydr. Res.*, 120, 17, 1983.

INDEX

A

α-D-Abep-(1→3)-α-D-Manp-(1→4)-L-Rha-(2←1)-α-D-Galp, 54
β-D-Abep-(1→3)-α-D-Manp-(1→4)-L-Rha-(2←1)-α-D-Galp, 55
α-Abep-(1→3)-α-D-Manp-(1→4)-α-L-Rhap-(1→3)-D-Gal, 72
α-L-Ascp-(1→3)-α-D-Manp-(1→4)-L-Rha-(2←1)-α-D-Galp, 55

C

α-Cyclodextrin, 121

D

2,3-Di-O-Me-α-L-Fucp-(1→4)-2,3-di-O-Me-α-L-Fucp-(1→3)-α-L-Rhap-(1→2)-6-deoxy-L-Tal, 82

F

α-L-Fucp-(1→2)-β-D-Galp-(1→4)-D-Glc-(3←1)-α-L-Fucp, 30
α-L-Fucp-(1→2)-β-D-Galp-(1→4)-D-GlcNAc-(3←1)-α-L-Fucp, 43
α-L-Fucp-(1→2)-β-D-Galp-(1→3)-D-GlcNAc-(3←1)-α-D-Galp, 36
α-L-Fucp-(1→2)-β-D-Galp-(1→3)-D-GlcNAc-(4←1)-α-D-Galp, 36
α-L-Fucp-(1→2)-β-D-Galp-(1→4)-D-GlcNAc-(3←1)-α-D-Galp, 39
α-L-Fucp-(1→2)-β-D-Galp-(1→4)-D-GlcNAc-(4←1)-α-D-Galp, 39
α-L-Fucp-(1→2)-β-D-Galp-(1→3)-D-GlcNAc-(3←1)-α-D-GalpNAc, 37
α-L-Fucp-(1→2)-β-D-Galp-(1→4)-D-GlcNAc-(3←1)-α-D-GalpNAc, 40
α-L-Fucp-(1→2)-β-D-Galp-(1→3)-β-D-GlcpNAc-(1→3)-β-D-Galp-(1→4)-D-Glc-(4←1)-α-L-Fucp, 124
α-L-Fucp-(1→4)-D-GlcNAc-(3←1)-α-L-Fucp-(1→2)-β-D-Galp, 42
α-L-Fucp-(1→3)-β-D-GlcpNAc-(1→2)-D-Man-(4←1)-β-D-Galp, 46
α-L-Fucp-(1→3)-β-D-GlcpNAc-(1→2)-α-D-Manp-(1→6)-D-Man-(3←1)-α-L-Fucp-(1→3)-β-D-GlcpNAc-(1→2)-α-D-Manp, 145

G

β-D-Galp-(1→4)-α-L-Fucp-(1→4)-α-L-Fucp-(1→3)-D-Glc, 29
β-D-Galp-(1→4)-β-L-Fucp-(1→4)-α-L-Fucp-(1→3)-D-Glc, 29
β-D-Galp-(1→6)-D-Gal-(2←1)-β-D-Glcp-(1→2)-β-D-Glcp, 79
β-D-Galp-(1→6)-β-D-Galp-(1→6)-β-D-Galp-(1→6)-3-deoxy-3-fluoro-D-Gal, 81
α-D-Galp-(1→6)-β-D-Galp-(1→6)-β-D-Galp-(1→6)-D-Gal, 74
β-D-Galp-(1→3)-β-D-Galp-(1→3)-β-D-Galp-(1→3)-D-Gal, 73
β-D-Galp-(1→6)-β-D-Galp-(1→6)-β-D-Galp-(1→6)-D-Gal, 74
β-D-Galp-(1→3)-β-D-Galp-(1→3)-β-D-Galp-(1→3)-β-D-Galp-(1→3)-D-Gal, 113
β-D-Galp-(1→6)-β-D-Galp-(1→6)-β-D-Galp-(1→6)-β-D-Galp-(1→6)-D-Gal, 114
β-D-Galp-(1→6)-β-D-Galp-(1→6)-β-D-Galp-(1→6)-β-D-Galp-(1→6)-β-D-Galp-(1→6)-D-Gal, 139
β-D-Galp-(1→3)-β-D-Galp-(1→3)-β-D-Galp-(1→3)-β-D-Galp-(1→3)-β-D-Galp-(1→3)-β-D-Galp-(1→3)-D-Gal, 149
β-D-Galp-(1→6)-α-D-Galp-(1→6)-α-D-Galp-(1→5)-D-Glc, 17
α-D-Galp-(1→6)-α-D-Galp-(1→6)-α-D-Glcp-(1→2)-D-Fru, 85
β-D-Galp-(1→6)-α-D-Galp-(1→6)-α-D-Glcp-(1→2)-D-Fru, 85
β-D-Galp-(1→3)-β-D-GalpNAc-(1→3)-β-D-Galp-(1→4)-D-Glc, 24
β-D-Galp-(1→3)-β-D-GalpNAc-(1→4)-β-D-Galp-(1→4)-D-Glc, 25
β-D-Galp-(1→3)-β-D-GalpNAc-(1→4)-β-D-Galp-(1→4)-D-Glc-(3←2)-α-Neu5Ac, 93
β-D-Galp-(1→3)-β-D-GalpN-(1→3)-β-D-Galp-(1→4)-D-Glc, 24
β-D-Galp-(1→3)-β-D-GalpN-(1→4)-β-D-Galp-(1→4)-D-Glc, 25
β-D-Galp-(1→4)-D-GlcNAc-(6←1)-β-D-Galp-(1→4)-β-D-Glcp, 41
β-D-Galp-(1→3)-β-D-GlcpNAc-(1→3)-D-Gal-(6←1)-β-D-Galp-(1→4)-β-D-GlcpNAc, 115
β-D-Galp-(1→4)-β-D-GlcpNAc-(1→3)-D-Gal-(6←1)-β-D-Galp-(1→4)-β-D-GlcpNAc, 116
β-D-Galp-(1→3)-β-D-GlcpNAc-(1→3)-D-Gal-(6←1)-β-D-GlcpNAc, 80
β-D-Galp-(1→4)-β-D-GlcpNAc-(1→6)-D-GalNAc-(3←1)-β-D-GlcpNAc, 80
β-D-Galp-(1→3)-β-D-GlcpNAc-(1→3)-β-D-Galp-(1→4)-D-Glc, 18
β-D-Galp-(1→4)-β-D-GlcpNAc-β-D-Galp-(1→4)-D-Glc, 21
β-D-Galp-(1→4)-β-D-GlcpNAc-(1→3)-β-D-Galp-(1→4)-D-Glc, 19
β-D-Galp-(1→4)-β-D-GlcpNAc-(1→6)-β-D-Galp-(1→4)-D-Glc-(3←1)-β-D-Galp-(1→4)-β-D-GlcpNAc, 125
β-D-Galp-(1→4)-β-D-GlcpNAc-(1→3)-β-D-Galp-(1→4)-D-GlcNAc, 38
β-D-Galp-(1→4)-β-D-GlcpNAc-(1→3)-β-D-Galp-(1→4)-β-D-GlcpNAc-(1→3)-β-D-Galp-(1→4)-D-GlcNAc, 127

β-D-Galp-(1→4)-β-D-GlcpNAc-(1→2)-D-Man-(4←1)-β-D-Galp-(1→4)-β-D-GlcpNAc, 103
β-D-Galp-(1→4)-β-D-GlcpNAc-(1→2)-D-Man-(6←1)-β-D-Galp-(1→4)-β-D-GlcpNAc, 104
β-D-Galp-(1→4)-β-D-GlcpNAc-(1→3)-α-D-Manp-(1→2)-D-Man-(6←1)-β-D-Galp-(1→4)-β-D-GlcpNAc, 130
β-D-Galp-(1→4)-β-D-GlcpNAc-(1→2)-α-D-Manp-(1→3)-D-Man-(6←1)-β-D-Galp-(1→4)-β-D-GlcpNAc-(1→2)-α-D-Manp, 160
β-D-Galp-(1→4)-β-D-GlcpNAc-(1→2)-α-D-Manp-(1→6)-D-Man-(3←1)-β-D-Galp-(1→4)-β-D-GlcpNAc-(1→2)-α-D-Manp, 145
β-D-Galp-(1→4)-β-D-GlcpNAc-(1→4)-α-D-Manp-(1→3)-D-Man-β-D-Galp-(1→4)-β-D-GlcpNAc-(1→2)-α-D-Manp-(1→6), 161
β-D-Galp-(1→4)-β-D-GlcpNAc-(1→2)-α-D-Manp-(1→6)-D-Man-(4←1)-β-D-GlcpNAc-β-D-Galp-(1→4)-β-D-GlcpNAc-(1→2)-α-D-Manp-(1→3), 155
β-D-Galp-(1→4)-β-D-GlcpNAc-(1→2)-α-D-Manp-(1→6)-D-Man-(3←1)-β-D-GlcpNAc-(1→2)-α-D-Manp-(3←1)-α-L-Fucp-β-D-Galp-(1→4), 159
β-D-Galp-(1→4)-β-D-GlcpNAc-(1→2)-α-D-Manp-(1→3)-D-Man-(6←1)-α-D-Manp, 106
β-D-Galp-(1→4)-β-D-GlcpNAc-(1→4)-α-D-Manp-(1→3)-D-Man-(6←1)-α-D-Manp, 107
β-D-Galp-(1→4)-β-D-GlcpNAc-(1→2)-α-D-Manp-(1→3)-D-Man-(6←1)-α-D-Manp-(4←1)-β-D-Galp-(1→4)-β-D-GlcpNAc, 146
β-D-Galp-(1→4)-β-D-GlcpNAc-(1→2)-α-D-Manp-(1→3)-D-Man-(6←1)-α-D-Manp-(6←1)-β-D-Galp-(1→4)-β-D-GlcpNAc, 166
β-D-Galp-(1→4)-β-D-GlcpNAc-(1→2)-β-D-Manp-(1→3)-D-Man-(6←1)-α-D-Manp-(4←1)-β-D-Galp-(1→4)-β-D-GlcpNAc, 148
β-D-Galp-(1→4)-β-D-GlcpNAc-(1→2)-α-D-Manp-(1→3)-β-D-Manp-(1→4)-D-GlcNAc, 95
β-D-Galp-(1→4)-β-D-GlcpNAc-(1→2)-α-D-Manp-(1→6)-β-D-Manp-(1→4)-D-GlcNAc, 96
β-D-Galp-(1→4)-β-D-GlcpNAc-(1→2)-α-D-Manp-(1→6)-β-D-Manp-(1→4)-D-GlcNAc-(3←1)-β-D-Galp-(1→4)-β-D-GlcpNAc-(1→2)-α-D-Manp, 154
β-D-Galp-(1→4)-β-D-GlcpNAc-(1→2)-α-D-Manp-(1→3)-β-D-Manp-(1→4)-β-D-GlcpNAc-(1→4)-D-GlcNAc-(6←1)-β-D-Galp-(1→4)-β-D-GlcpNAc-(1→2)-α-D-Manp, 158
β-D-Galp-(1→4)-β-D-GlcpN-(1→4)-β-D-Galp-(1→4)-D-Glc, 20
β-D-Galp-(1→4)-α-D-ManpNAc-(1→3)-β-D-Galp-(1→4)-D-Glc, 22
β-D-Galp-(1→4)-α-D-ManpN-(1→3)-β-D-Galp-(1→4)-D-Glc, 22
α-D-Galp-(1→2)-α-D-Manp-(1→4)-L-Rha-(3←1)-α-D-Parp, 53
α-D-Galp-(1→2)-α-D-Manp-(1→4)-L-Rha-(3←1)-β-D-Parp, 53
α-D-Galp-(1→2)-α-D-Manp-(1→4)-L-Rha-(3←1)-3,4,6-trideoxy-α-D-erythro-Hexp, 56
α-D-Galp-(1→2)-α-D-Manp-(1→4)-L-Rha-(3←1)-3,4,6-trideoxy-β-D-erythro-Hexp, 57
α-D-Galp-(1→2)-α-D-Manp-(1→4)-L-Rha-(3←1)-2,3,6-trideoxy-α-D-threo-Hexp, 56
α-D-Galp-(1→2)-α-D-Manp-(1→4)-L-Rha-(3←1)-α-D-Tyvp, 54
β-D-GalpNAc-(1→3)-α-D-Galp-(1→4)-β-D-Galp-(1→4)-D-Glc, 23
α-D-GalpNAc-(1→3)-β-D-GalpNAc-(1→3)-α-D-Galp-(1→4)-β-D-Galp-(1→4)-D-Glc, 92
α-D-GalpNAc-(1→3)-α-D-GalpNAc-(1→4)-D-Glc-(4←1)-α-D-GlcpNAc, 28
α-D-GalpNAc-(1→3)-α-D-GalpNAc-(1→4)-α-D-Glcp-(1→4)-D-Gal-(4←1)-α-D-GlcpNAc, 113
β-D-GalpNAc-(1→2)-β-L-Rhap-(1→2)-L-Rha-(4←1)-β-L-Rhap, 61
β-D-GalpNAc-(1→2)-β-L-Rhap-(1→4)-L-Rha-(4←1)-β-L-Rhap, 62
β-D-GalpN-(1→3)-α-D-Galp-(1→4)-β-D-Galp-(1→4)-D-Glc, 23
β-D-GalpN-(1→4)-β-D-Galp-(1→4)-D-Glc-(3←1)-β-D-GlcpN, 27
β-D-GalpN-(1→4)-β-D-Galp-(1→4)-D-Glc-(3←2)-α-Neu5Ac, 27
β-D-Galp-(1→3)-α-L-Rhap-(1→3)-α-L-Rhap-(1→2)-L-Rha, 58
β-D-GlcpA-(1→4)-α-D-GlcpN-(1→4)-α-L-IdopA-(1→4)-D-GlcN, 35
α-D-GlcpA-(1→6)-α-D-Glcp-(1→2)-L-Rha-(3←1)-α-L-Rhap, 65
β-D-GlcpA-(1→6)-α-D-Glcp-(1→2)-L-Rha-(3←1)-α-L-Rhap, 66
β-D-Glcp-(1→4)-2,6-dideoxy-2-I-α-D-Altp-(1→3)-2,6-dideoxy-α-D-ribo-Hexp-(1→4)-2,6-dideoxy-D-ribo-Hexp, 83
β-D-Glcp-(1→4)-2,6-dideoxy-α-D-ribo-Hexp-(1→4)-2,6-dideoxy-α-D-ribo-Hexp-(1→4)-2,6-dideoxy-D-ribo-Hexp, 84
α-D-Glcp-(1→4)-D-Gal-(3←1)-β-D-Manp-(1→4)-α-L-Rhap, 76
α-D-Glcp-(1→6)-D-Gal-(3←1)-α-D-Manp-(1→4)-α-L-Rhap, 77
α-D-Glcp-(1→6)-D-Gal-(3←1)-β-D-Manp-(1→4)-α-L-Rhap, 77
α-D-Glcp-(1→6)-D-Gal-(3←1)-β-D-Manp-(1→4)-β-L-Rhap, 78
β-D-Glcp-(1→6)-β-D-Galp-(1→6)-D-Gal-(2←1)-β-D-Glcp, 79
β-D-Glcp-(1→2)-β-D-Galp-(1→2)-β-D-Galp-(1→6)-D-Gal, 73
α-D-Glcp-(1→4)-D-Gal-(3←1)-α-D-Tyvp-(1→3)-β-D-Manp-(1→4)-α-L-Rhap, 115
β-D-Glcp-(1→2)-D-Glc-(6←1)-β-D-Glcp-α-L-Rhap-(1→4), 31
β-D-Glcp-(1→2)-β-D-Glcp-(1→4)-D-Gal-(3←1)-β-D-Xylp, 70

α-D-Glcp-(1→6)-α-D-Glcp-(1→4)-D-Glc-(6←1)-α-D-Glcp, 15
β-D-Glcp-(1→3)-β-D-Glcp-(1→3)-D-Glc-(6←1)-β-D-Glcp, 9, 16
β-D-Glcp-(1→3)-β-D-Glcp-(1→6)-D-Glc-(6←1)-β-D-Glcp, 14
β-D-Glcp-(1→4)-β-D-Glcp-(1→4)-D-Glc-(1→6)-β-D-Glcp, 12
β-D-Glcp-(1→6)-α-D-Glcp-(1→4)-D-Glc-(6←1)-β-D-Glcp, 15
β-D-Glcp-(1→3)-β-D-Glcp-(1→6)-D-Glc-(6←1)-β-D-Glcp-(1→6)-β-D-Glcp-(6-1-β-D-Glcp-β-D-Glcp-(1→3), 142
α-D-Glcp-(1→4)-α-D-Glcp-(1→4)-α-D-Glcp-(1→4)-D-Glc, 5
α-D-Glcp-(1→4)-β-D-Glcp-(1→4)-α-D-Glcp-(1→4)-D-Glc, 5
α-D-Glcp-(1→4)-β-D-Glcp-(1→6)-α-D-Glcp-(1→4)-D-Glc, 6
α-D-Glcp-(1→6)-α-D-Glcp-(1→6)-α-D-Glcp-(1→6)-D-Glc, 7
β-D-Glcp-(1→2)-β-D-Glcp-(1→2)-β-D-Glcp-(1→2)-D-Glc, 8
β-D-Glcp-(1→3)-β-D-Glcp-(1→3)-β-D-Glcp-(1→6)-D-Glc, 12
β-D-Glcp-(1→3)-β-D-Glcp-(1→4)-β-D-Glcp-(1→4)-D-Glc, 10
β-D-Glcp-(1→4)-β-D-Glcp-(1→3)-β-D-Glcp-(1→4)-D-Glc, 10
β-D-Glcp-(1→4)-β-D-Glcp-(1→4)-β-D-Glcp-(1→3)-D-Glc, 9
β-D-Glcp-(1→4)-β-D-Glcp-(1→4)-β-D-Glcp-(1→4)-D-Glc, 11
β-D-Glcp-(1→6)-β-D-Glcp-(1→3)-β-D-Glcp-(1→3)-D-Glc, 8
β-D-Glcp-(1→6)-β-D-Glcp-(1→6)-β-D-Glcp-(1→6)-D-Glc, 13
α-D-Glcp-(1→3)-α-D-Glcp-(1→6)-α-D-Glcp-(1→6)-D-Glc-(3←1)-α-D-Glcp-(1→3)-α-D-Glcp-(1→6)-α-D-Glcp, 143
β-D-Glcp-(1→3)-β-D-Glcp-(1→3)-β-D-Glcp-(1→6)-D-Glc-(6←1)-β-D-Glcp-(1→6)-β-D-Glcp-(3←1)-β-D-Glcp-β-D-Glcp-(1→6), 153
α-D-Glcp-(1→6)-α-D-Glcp-(1→6)-α-D-Glcp-(1→6)-α-Glcp-(1→6)-D-Glc, 90
β-D-Glcp-(1→2)-β-D-Glcp-(1→2)-β-D-Glcp-(1→2)-β-D-Glcp-(1→2)-D-Glc, 90
β-D-Glcp-(1→4)-β-D-Glcp-(1→3)-β-D-Glcp-(1→4)-β-D-Glcp-(1→4)-D-Glc, 91
β-D-Glcp-(1→4)-β-D-Glcp-(1→4)-β-D-Glcp-(1→4)-β-D-Glcp-(1→4)-D-Glc, 91
β-D-Glcp-(1→6)-β-D-Glcp-(1→6)-β-D-Glcp-(1→6)-β-D-Glcp-(1→6)-D-Glc, 92
α-D-Glcp-(1→4)-α-D-Glcp-(1→4)-α-D-Glcp-(1→4)-α-D-Glcp-(1→4)-α-D-Glcp-(1→4)-D-Glc, 119
α-D-Glcp-(1→4)-β-D-Glcp-(1→4)-α-D-Glcp-(1→4)-α-D-Glcp-(1→4)-α-D-Glcp-(1→4)-D-Glc, 120
α-D-Glcp-(1→6)-α-D-Glcp-(1→6)-α-D-Glcp-(1→6)-α-D-Glcp-(1→6)-α-D-Glcp-(1→6)-D-Glc, 122
β-D-Glcp-(1→6)-β-D-Glcp-(1→6)-β-D-Glcp-(1→6)-β-D-Glcp-(1→6)-β-D-Glcp-(1→6)-D-Glc, 123
α-D-Glcp-(1→6)-α-D-Glcp-(1→6)-α-D-Glcp-(1→6)-α-D-Glcp-(1→6)-α-D-Glcp-(1→6)-α-D-Glcp-(1→6)-D-Glc, 152
α-D-Glcp-(1→2)-α-D-Glcp-(1→3)-α-D-Glcp-(1→3)-D-Man, 44
α-D-Glcp-(1→2)-α-D-Glcp-(1→3)-α-D-Glcp-(1→3)-α-D-Manp-(1→2)-α-D-Manp-(1→2)-D-Man, 129
α-D-Glcp-(1→4)-α-D-Glcp-(1→3)-L-Rha-(4←1)-β-D-ManpNAc, 69
α-D-Glcp-(1→4)-α-D-Glcp-(1→3)-α-L-Rhap-(1→3)-D-ManNAc, 52
α-D-Glcp-(1→4)-β-D-Glcp-(1→3)-L-Rhp-(4←1)-β-D-ManpN, 70
β-D-Glcp-(1→2)-D-Glc-(4←1)-α-L-Rhap-β-D-Glcp-(1→6), 31
β-D-GlcpNAc-(1→6)-β-D-Galp-(1→3)-D-GalNAc-(6←1)-β-D-GlcpNAc, 81
β-D-GlcpNAc-(1→3)-β-D-Galp-(1→4)-D-Glc-(6←1)-β-D-GlcpNAc, 26
β-D-GlcpNAc-(1→3)-α-D-Galp-(1→4)-L-Rha-(4←1)-β-D-Manp, 63
β-D-GlcpNAc-(1→3)-β-D-Galp-(1→4)-L-Rha-(4←1)-β-D-Manp, 64
β-D-GlcpNAc-(1→4)-β-D-GlcpNAc-(1→6)-β-D-GlcpNAc-(1→4)-D-GlcNAc, 32
β-D-GlcpNAc-(1→4)-D-Man-(3←1)-α-D-Manp-α-D-Manp-(1→6), 51
β-D-GlcpNAc-(1→4)-β-D-Manp-(1→4)-D-GlcNAc-(3←1)-α-D-Manp-α-D-Manp-(1→6), 99
β-D-GlcpNAc-(1→2)-α-D-Manp-(1→6)-D-Man-(6←1)-β-D-GlcpNAc, 48
β-D-GlcpNAc-(1→2)-α-D-Manp-(1→3)-D-Man-(4←1)-β-D-GlcpNAc-β-D-GlcpNAc-(1→2)-α-D-Manp-(1→6), 136
β-D-GlcpNAc-(1→2)-α-D-Manp-(1→3)-D-Man-(6←1)-α-D-GlcpNAc-(1→2)-α-D-Manp, 108
β-D-GlcpNAc-(1→2)-α-D-Manp-(1→3)-D-Man-(6←1)-α-D-Manp, 50
β-D-GlcpNAc-(1→2)-α-D-Manp-(1→6)-D-Man-(3←1)-α-D-Manp, 50
β-D-GlcpNAc-(1→2)-α-D-Manp-(1→3)-D-Man-(6←1)-α-D-Manp-(2←1)-β-D-GlcpNAc-β-D-GlcpNAc-(1→4), 135
β-D-GlcpNAc-(1→2)-α-D-Manp-(1→3)-β-D-Manp-(1→4)-D-GlcNAc-(6←1)-α-D-Manp, 98
β-D-GlcpNAc-(1→2)-α-D-Manp-(1→6)-β-D-Manp-(1→4)-D-GlcNAc-(3←1)-α-D-Manp, 97
β-D-GlcpNAc-(1→2)-α-L-Rhap-(1→2)-L-Rha-(3←1)-α-D-Glcp, 68
β-D-GlcpNAc-(1→2)-α-L-Rhap-(1→2)-L-Rha-(3←1)-β-D-Glcp, 69
β-D-GlcpNAc-(1→2)-α-L-Rhap-(1→2)-L-Rha-

(3←1)-α-D-Glcp-(3←1)-α-D-Glcp, 112
β-D-GlcpNAc-(1→2)-α-L-Rhap-(1→2)-α-L-Rhap-(1→3)-L-Rha, 60
β-D-GlcpNAc-(1→2)-α-L-Rhap-(1→2)-α-L-Rhap-(1→3)-α-L-Rhap-(1→3)-β-D-GlcpNAc-(1→2)-α-L-Rhap-(1→2)-α-L-Rhap-(1→3)-L-Rha, 156
α-D-GlcpN-(1→3)-α-D-Galp-(1→4)-L-Rhap-(4←1)-β-D-Manp, 62
β-D-GlcpN-(1→4)-β-D-GlcpA-(1→4)-α-D-GlcpN-(1→4)-α-L-IdopA-(1→4)-D-GlcN, 101
α-D-GlcpN-(1→6)-α-D-GlcpN-(1→6)-α-D-GlcpN-(1→6)-D-GlcN, 31
β-D-GlcpN-(1→2)-α-D-Manp-(1→3)-β-D-Manp-(1→4)-D-Glcn-(6←1)-β-D-GlcpN-(1→2)-α-D-Manp, 126
α-D-GlcpNSO$_3$(6-OSO$_3$)-(1→4)-β-D-GlcpA-(1→4)-α-D-GlcpNSO$_3$)-(1→4)-α-L-IdopA (2-OSO$_3$)-(1→4)-D-GlcNSO, 99
α-D-GlcpNSO$_3$-(6-OSO$_3$)-(1→4)-β-D-GlcpA-(1→4)-α-D-GlcpNSO$_3$-(1→4)-α-L-IdopA (2-OSO$_3$ (6-OSO$_3$), 100
β-D-Glcp-(1→4)-α-L-Rhap-(1→3)-D-Gal-(6←1)-α-D-Glcp, 76
β-D-Glcp-(1→4)-β-L-Rhap-(1→3)-D-Gal-(6←1)-α-D-Glcp, 78
β-D-Glcp-(1→4)-α-L-Rhap-(1→3)-β-D-Galp-(1→6)-D-Glc, 28
β-D-Glcp-(1→4)-α-L-Rhap-(1→3)-β-D-Galp-(1→6)-β-D-Glcp-(1→4)-α-L-Rhap-(1→3)-D-Gal, 137
α-D-Glcp-(1→3)-α-L-Rhap-(1→2)-L-Rha-(3←1)-α-D-Glcp, 67
α-D-Glcp-(1→3)-α-L-Rhap-(1→2)-L-Rha-(2←1)-β-D-GlcpNAc, 59
α-D-Glcp-(1→3)-α-L-Rhap-(1→3)-L-Rha-(2←1)-α-L-Rhap, 61

K

α-KDO-(2→4)-α-KDO-(2→6)-β-D-GlcpN-(1→6)-D-GlcN, 33

M

α-D-Manp-(1→3)-D-Man-(6←1)-β-D-GlcpNAc-(1→2)-α-D-Manp, 49
α-D-Manp-(1→2)-D-Man-(4←1)-α-D-Manp-α-D-Manp-(1→6), 51
α-D-Manp-(1→3)-β-D-Manp-(1→4)-D-GlcNAc-(6←1)-α-D-Manp, 34
α-D-Manp-(1→3)-β-D-Manp-(1→4)-β-D-GlcpNAc-(1→4)-D-GlcNAc, 32
α-D-Manp-(1→3)-β-D-Manp-(1→4)-β-D-GlcpNAc-(1→4)-D-GlcNAc-(1→4)-D-GlcNAc-(6←1)-α-D-Manp, 94
α-D-Manp-(1→3)-α-D-Manp-(1→6)-D-Man-(3←1)-α-D-Manp, 50
α-D-Manp-(1→3)-α-D-Manp-(1→6)-D-Man-(6←1)-α-D-Manp, 49
α-D-Manp-(1→2)-α-D-Manp-(1→2)-D-Man-(4←1)-α-Manp-(1→2)-α-D-Manp, 105
α-D-Manp-(1→2)-α-D-Manp-(1→2)-D-Man-(6←1)-α-D-Manp-(1→2)-α-D-Manp, 106
α-D-Manp-(1→2)-α-D-Manp-(1→3)-D-Man-(6←1)-α-D-Manp-(1→2)-α-D-Manp, 108
α-D-Manp-(1→2)-α-D-Manp-(1→6)-D-Man-(3←1)-α-D-Manp-(6←1)-α-D-Manp, 109
α-D-Manp-(1→3)-α-D-Manp-(1→3)-D-Man-(6←1)-α-D-Manp-(6←1)-α-D-Manp, 110
α-D-Manp-(1→3)-α-D-Manp-(1→6)-D-Man-(3←1)-α-D-Manp-(6←1)-α-D-Manp, 111
α-D-Manp-(1→2)-α-D-Manp-(1→3)-D-Man-(6←1)-α-D-Manp-(2←1)-α-D-Manp-α-D-Manp-(1→6), 132
α-D-Manp-(1→3)-α-D-Manp-(1→3)-D-Man-(6←1)-α-D-Manp-(1→2)-α-D-Manp-(6←1)-α-D-Manp, 134
α-D-Manp-(1→3)-α-D-Manp-(1→6)-D-Man-(3←1)-α-D-Manp-(1→2)-α-D-Manp-(1→3)-(6←1)-α-D-Manp, 133
α-D-Manp-(1→2)-α-D-Manp-(1→3)-β-D-Manp-(1→4)-D-GlcNAc, 33
α-D-Manp-(1→3)-α-D-Manp-(1→6)-β-D-Manp-(1→4)-D-GlcNAc, 34
α-D-Manp-(1→2)-α-D-Manp-(1→2)-α-D-Manp-(1→2)-D-Man, 46
α-D-Manp-(1→2)-α-D-Manp-(1→6)-α-D-Manp-(1→6)-D-Man, 47
α-D-Manp-(1→2)-β-D-Manp-(1→6)-α-D-Manp-(1→6)-D-Man, 48
α-D-Manp-(1→3)-α-D-Manp-(1→3)-α-D-Manp-(1→3)-D-Man, 47
α-D-Manp-(1→2)-α-D-Manp-(1→3)-α-D-Manp-(1→6)-D-Man-(6←1)-α-D-Manp-(1→2)-α-D-Manp, 131
α-D-Manp-(1→3)-α-D-Manp-(1→2)-α-D-Manp-(1→2)-α-D-Manp-(1→2)-D-Man, 102
α-D-Manp-(1→3)-α-D-Manp-(1→3)-α-D-Manp-(1→3)-α-D-Manp-(1→3)-D-Man, 102
α-D-Manp-(1→2)-α-D-Manp-(1→2)-α-D-Manp-(1→2)-α-D-Manp-(1→2)-α-D-Manp-(1→2)-D-Man, 128
α-D-Manp-(1→3)-α-D-Manp-(1→3)-α-D-Manp-(1→3)-α-D-Manp-(1→3)-α-D-Manp-(1→3)-D-Man, 130
β-D-Manp-(1→4)-α-L-Rhap-(1→3)-β-D-Galp-(1→6)-β-D-Manp-(1→4)-α-L-Rhap-(1→3)-D-Gal, 138
β-D-Manp-(1→4)-α-L-Rhap-(1→3)-β-D-Galp-(1→6)-β-D-Manp-(1→4)-α-L-Rhap-(1→3)-β-D-Galp-(1→6)-β-D-Manp-(1→4)-α-L-Rhap-(1→-D-Gal, 162
2-O-Me-α-L-Fucp-(1→4)-2-O-Me-α-L-Fucp-(1→3)-α-L-Rhap-(1→2)-6-deoxy-L-Tal, 82
3-O-Me-β-L-Xylp-(1→4)-α-L-Rhap-(1→4)-α-L-

Rhap-(1→2)-L-Rha,58

N

α-D-Neu5Ac-(2→6)-β-D-Galp-(1→4)-β-D-GlcpNAc-(1→2)-D-Man, 44
β-D-Neu5Ac-(2→6)-β-D-Galp-(1→4)-β-D-GlcpNAc-(1→2)-D-Man, 45
α-Neu5Ac-(2→6)-β-D-Galp-(1→4)-β-D-GlcpNAc-(1→2)-α-D-Manp-(1→6)-β-D-Manp-(1→4)-β-D-GlcpNAc-(1→4)-D-GlcNAc, 144
α-Neu5Ac-(2→6)-β-D-Galp-(1→4)-β-D-GlcpNAc-(1→2)-α-D-Manp-(1→3)-β-D-Manp-(1→4)-β-D-GlcpNAc-(1→4)-D-GlcNAc-(6←1)-α-Neu5Ac-(2→6)-β-D-Galp-(1→4)-β-D-Glcp, 164

P

Parp-3,6-Dideoxy-D-ribo-hexopyranosyl, 71
α-Parp-(1→3)-α-D-Manp-(1→4)-α-L-Rhap-(1→3)-D-Gal, 71
β-Parp-(1→3)-α-D-Manp-(1→4)-α-L-Rhap-(1→3)-D-Gal, 71

R

α-L-Rhap-(1→3)-α-L-Rhap-(1→3)-β-D-GlcpNAc-(1→2)-L-Rha, 52
α-L-Rhap-(1→2)-α-L-Rhap-(1→3)-α-L-Rhap-(1→3)-D-GlcNAc, 35
α-L-Rhap-(1→2)-α-L-Rhap-(1→3)-α-L-Rhap-(1→2)-L-Rha, 57

T

Tyvp-3,6-Dideoxy-D-arabino-hexopyranosyl, 72
α-Tyvp-(1→3)-α-D-Manp-(1→4)-α-L-Rhap-(1→3)-D-Gal, 72

X

α-D-Xylp-(1→4)-β-D-Xylp-(1→4)-β-D-Xylp-(1→4)-D-Xyl, 2
β-D-Xylp-(1→3)-β-D-Xylp-(1→3)-β-D-Xylp-(1→3)-D-Xyl, 2
β-D-Xylp-(1→4)-β-D-Xylp-(1→4)-β-D-Xylp-(1→4)-D-Xyl, 3
β-D-Xylp-(1→3)-β-D-Xylp-(1→4)-(4←1)-β-D-Xylp-D-Xyl-(2←1)-4-O-Me-α-D-GlcpA, 89
β-D-Xylp-(1→3)-β-D-Xylp-(1→4)-(4←1)-β-D-Xylp-D-Xyl-(2←1)-4-O-Me-β-D-GlcpA, 89
α-D-Xylp-(1→4)-β-D-Xylp-(1→4)-β-D-Xylp-(1→4)-β-D-Xylp-(1→4)-D-Xyl, 88
β-D-Xylp-(1→4)-β-D-Xylp-(1→4)-β-D-Xylp-(1→4)-β-D-Xylp-(1→4)-D-Xyl, 88
α-D-Xylp-(1→4)-β-D-Xylp-(1→4)-β-D-Xylp-(1→4)-β-D-Xylp-(1→4)-β-D-Xylp-(1→4)-D-Xyl, 118
β-D-Xylp-(1→4)-β-D-Xylp-(1→4)-β-D-Xylp-(1→4)-β-D-Xylp-(1→4)-β-D-Xylp-(1→4)-D-Xyl, 118
β-D-Xylp-(1→3)-β-D-Xylp-(1→4)-D-Xyl-(4←1)-β-D-Xylp, 4